Technical Writing

Technical Writing

2nd Edition

by Sheryl Lindsell-Roberts, MA

Technical Writing For Dummies®, 2nd Edition

Published by

Wiley Publishing, Inc.

111 River St.

Hoboken, NJ 07030-5774

www.wiley.com

For general information on our other products and services, please contact our Customer Care Department within the U.S. at 877-762-2974, outside the U.S. at 317-572-3993, or fax 317-572-4002. For technical support, please visit https://hub.wiley.com/community/support/dummies.

Wiley publishes in a variety of print and electronic formats and by print-on-demand. Some material included with standard print versions of this book may not be included in e-books or in print-on-demand. If this book refers to media such as a CD or DVD that is not included in the version you purchased, you may download this material at http://booksupport.wiley.com. For more information about Wiley products, visit www.wiley.com.

Library of Congress Control Number: 2023937438

ISBN 978-1-394-17675-5 (pbk); ISBN 978-1-394-17676-2 (ebk); ISBN 978-1-394-17677-9 (ebk)

SKY10047806_051523

Contents at a Glance

Table of Contents

Introduction

As a technical communicator, I am an enabler of information.

—SUYOG KETKAR, CERTIFIED TECHNICAL WRITER

Welcome to the second edition of *Technical Writing For Dummies*, which will propel you into the exciting worlds of eLearning, collaboration tools, videoconferencing, streaming, simulations, surfing, artificial intelligence (AI), virtual reality (VR), search engine optimization (SEO), user experience (UX) writing, a zoom into the metaverse, and much more!

About This Book

To make the content easy to find and read, I divided it into six parts:

- **Part 1: What It Takes to Write Technical Docs.** This part takes you through the gratification of being a technical writer and discusses putting together a team.
- **Part 2: The Write Stuff.** This part introduces you to the Technical Writing Brief to find out all you can about your learners. Following that, you discover how to write a draft, design for visual impact, hone the tone, and proofread and edit. This is the meat of the book, so you'll get the most out of it by reading these chapters in sequential order.
- **Part 3: Frequently Written Docs.** Here, you find guidelines for writing whiz-bang user manuals (and more), plus scripting for streaming and simulations, abstracts, spec sheets, questionnaires, technical presentations, and executive summaries.
- **Part 4: Tech Tools.** No technical writing book would be complete without focusing on the power of people, the computer, and the Internet. This part goes into details about team collaboration, videoconferencing, eLearning, doing advanced online searches, and protecting intellectual property.
- **Part 5: The Part of Tens.** This part is a *Dummies* classic. It includes a potpourri of tips on writing whitepapers and journal articles, as well as some frustrations of technical writers.
- **Appendixes.** The appendixes round out this book with sections on punctuation, grammar, abbreviations and metrics, tech terms, and a blank copy of the Technical Writing Brief.

Beyond the words on the pages, I practice what I preach and teach. I wrote this book in conversational language, much like I'd talk to you. I hope you'll learn by example and write your technical documents as if you're talking to your learners. This will make you and your writing relatable. Throughout this book, you see the following to emulate:

- Informative headlines and subheads
- Bulleted and numbered lists
- Straightforward, conversational language
- Short paragraphs
- Graphic elements (when they "speak" louder than words)

Foolish Assumptions

Before I began writing this book, I made some assumptions about you — the reader. (I don't normally make assumptions because we all know what happens when we ass-u-me.) However, I've thrown caution to the wind and guessed that you likely fit into one of these categories:

- You're an engineer, scientist, computer programmer, or IT specialist
- You work in some other technical or medical field
- You're a newbie writer or a seasoned technical writer
- You're a college student who's entering a technical field or becoming a tech writer
- You shake and grunt like an unbalanced clothes dryer when you're asked to write a technical document

Technical writers come from all walks of life: teachers, musicians, tradespeople, journalists, financial analysts, attorneys, scientists, researchers, and more. Technical writing services are sought worldwide and are in hot demand. So whether you're a technical person who finds that technical writing is something you must do to advance your career or you're a professional technical writer looking to fine-tune your skills, this book is invaluable to your professional growth and survival.

And that's a *sensible* assumption!

NOTE ABOUT GENDERS

The language of genders has evolved over time. Many of us have become accustomed to using she/her/hers for females and he/his/him for males. As gender vocabulary continues to evolve, it's proper to address a singular person as they, them, ze, or hir. Many people now put their gender preferences in the signature blocks of their emails. Inclusive language offers respect, safety, and belonging to all people. This book uses this inclusive approach.

Having said that, *The Chicago Manual of Style* (a staple reference for writers and editors since 1906) is watching the generic use of inclusive language, stating: "They and their have become common in informal usage, but neither is considered fully acceptable in formal writing." As you go through your technical writing projects, use good judgment and always consider your audience.

Icons Used in This Book

Scattered throughout this book are icons in the margins. They highlight valuable information that call for your attention.

SHERYL SAYS

Benefit from my experiences — the blissful, the painful, and everything in between.

TECHNICAL WRITING BRIEF

Don your Sherlock Holmes hat, scrutinize the Technical Writing Brief, and gather all the clues you can about your readers.

TIP

Find nifty tips for writing it so they'll read it. These may be time savers, frustration savers, lifesavers, or just about anything else.

REMEMBER

This is just what you'd expect. What else? — tidbits to remember.

WARNING

Ouch! Avoid these pitfalls to save yourself headaches, heartburn, embarrassments, or worse.

CROSS REFERENCE

Check out another section of the book for related content.

Beyond the Book

In addition to the material in the print or e-book you're reading right now, this product also comes with some goodies you can access on the web. Check out the free access-anywhere Cheat Sheet that includes tips and advice. To get this Cheat Sheet, simply go to `www.dummies.com` and type **Technical Writing For Dummies Cheat Sheet** in the Search box.

The cheat sheet is a blank Technical Writing Brief, which you can use to jump-start all writing projects. (A copy also appears in Chapter 3 with an in-depth explanation, as well as a blank paper copy in Appendix E.) Share the brief with your team and keep a copy on your computer, tablet, and smartphone for handy reference.

Where to Go from Here

I realize you won't read this book like a suspenseful mystery novel from cover to cover — but I strongly suggest that you read Part 2 (Chapters 3-7) sequentially. Good technical writing is a process of understanding your learners, writing the draft, designing for visual impact, honing the tone, and proofreading and editing. These chapters offer the foundation for a wide variety of technical documents, many of which appear in his book. After that, feel free to jump around to whatever topic interests you or applies to the writing challenges you face.

1 What It Takes to Write Technical Docs

IN THIS PART . . .

Develop a strategic approach to technical writing, produce impactful documents, learn how to create a portfolio, and discover tech writing career trajectories.

Put together the best team, develop your plan, choose the delivery method, and complete a production schedule and outline.

IN THIS CHAPTER

» Discovering who writes technical documents

» Understanding how business and technical documents differ

» Creating a portfolio and business cards

» Learning about different career trajectories

Chapter 1

Working as a Technical Writer

I didn't go to film school. My grandpa always says just watch a lot of movies. He didn't go to film school; he went to theatre school. It's interesting to learn about the technical side of it, but I think it's more important to learn about writing and working with actors.

—Gia Coppola, Granddaughter of Francis Ford Coppola

Although formal training in technical writing may be helpful, you don't *need* it any more than Francis Ford Coppola needed film school to become one of the most successful figures of Hollywood filmmaking. What you need is

- A love of learning
- An attention to detail
- A good command of the English language

- An understanding of how people use and process information
- The ability to manage tasks and work well as part of a team

If you arranged your alphabet soup into acronyms when you were a kid, you constantly asked "why" when people told you to stop asking questions, or you sent Santa lists with headings and subheadings, you're a natural-born technical writer.

Technical Writers Spring from All Walks of Life

People who write technical documents come from all walks of life — and most aren't technical writers per se. Here are some actual situations of people who were called upon to write technical documents in the course of their professions:

- **Computer programmer:** Octavia graduated with a degree in computer science and was hired as a software developer for a company. Several months later, the company felt a financial pinch and laid off the technical writers. Octavia had a big deliverable due in a few months, and her supervisor told her that she had to write a user manual. Sophomore English (which Octavia struggled through and loathed) didn't prepare her for this type of assignment. After all, Shakespeare wasn't a technical sort of guy. Poor Octavia had to muddle through writing the user manual and got gray hair prematurely.
- **Manufacturing specialist:** Bill worked for a manufacturing company for many years and developed a piece of equipment that was expected to revolutionize the industry. The equipment made its debut in Germany at the industry's largest conference. Bill's supervisor asked him to deliver a paper (the industry term for a making technical presentation) at the conference. The audience would consist of more than 200 high-level industry professionals. Not only did Bill fear the podium more than the dentist's drill, he didn't know how to prepare or deliver a technical paper — especially in a foreign country for an audience of this caliber.
- **Biotech scientist:** While working at a pharmaceutical company, Abdul had a major breakthrough on a treatment that promised to prevent baldness. The company president asked him to write an article for a major medical journal. Although Abdul was flattered by the president's request, he didn't know the first thing about writing or submitting a technical article.

BRIGHT AND EXCITING FUTURES FOR TECH WRITERS

The Bureau of Labor Statistics released its annual Occupational Handbook for Technical Writers. It predicts that job growth in this field is expected to outpace the national average for all other occupations. This is due in part to the growing high-tech and electronics industries that are embracing the value of superior-quality technical communications — paving the way to solve problems more quickly and easily by intersecting the human experience with the digital world.

- **Sales representative:** Lynette was a sales representative for a worldwide computer distributor. She'd often be away from home for weeks at a time. After 15 years as a road warrior, Lynette suffered from burnout. (She used to leave her picture on the fireplace mantle so that her family wouldn't forget what she looked like.) Lynette had been reading about the burgeoning field of tech writing. She called a local college, got all the literature, and decided to pursue a degree in technical writing.

SHERYL SAYS

Although I changed the names to protect the innocent, scenarios such as these are typical. Technical people who aren't trained writers are constantly asked to write technical documents. Their education and work experience rarely prepare them for this type of challenge. This book can help bridge the gap!

Documentation Is Part of Our Everyday Lives

Whether you realize it or not, documentation is part of our everyday lives — both personal and business. When you buy a new piece of electronics, it comes with instructions. When you buy a DYI (do-it-yourself) furniture kit, it comes with assembly instructions. When you get a prescription for medication, it comes with a leaflet on how often to take the medication and what the side effects may be. Documents are written for all of us, not just for computer geeks who assemble rockets or plasma generators. And it's not just the computer geeks who write technical documents — all technical people do at some point in their careers.

THE HUMBLE BEGINNINGS OF TECH WRITING

Technical writing as we know it today took root in World War II when the U.S. military persuaded "those who served" to write manuals to aid the war effort. The military needed to teach soldiers about weapons, transport vehicles, and other hardware. These "technical writers" had little or no training. They just sat down at their manual typewriters and banged out whatever made sense to them. I don't know whether it made sense to the poor soldiers trying to decipher their writing. But we did win the war!

Technical writing means different things to different people. It covers the fields of electronics, aircraft, computer manufacturing and software development, chemical, biotech, pharmaceuticals, health, and much more. It spans the public and private sectors as well as government and academic institutions.

Technical Writing Differs from Business Writing

Many people ask the difference between business writing and technical writing. The difference is analogous to apples and watermelon. For example, at the very core (pardon the pun), apples and watermelons are fruits. And at the very core, documents are words and graphics. Beyond the core, business and technical documents are different species.

Documents of the business kind

Emails are the crux of business writing and account for as much as 90 percent of all business communication. Other type of business writing include letters, reports, blog posts, articles, and more. One major difference between business and technical documents is that business documents are generally written by one person, often for a single learner or small, select group of learners. Following are some commonly written business documents:

- Agendas
- Emails
- Letters

- Meeting minutes
- Proposals

Documents of the technical kind

People in specialized fields write documents that relate to technical or complex subjects. Unlike business documents, technical documents are often a collaborative effort between a technical writer, UX writer, subject matter expert (SME), editor, and others. Technical documents are generally intended for a vast number of learners. Following are some commonly written technical documents. You find chapters about the specifics of writing each of these documents later in this book.

- Abstracts
- Articles for publication
- eLearning
- Executive summaries
- Functional and detail specifications
- Online help
- Questionnaires
- Reports
- User manuals

Assigning Responsibility for Technical Documents

The responsibility for writing technical documents depends on a company's structure and resources. Following are several ways that companies typically generate technical documents:

- **Technical gurus (engineers, software developers, and others) write their own documents.** Some of these people may have taken writing courses, but most have no training in writing a cohesive document. These "technical writers" often overlook steps in the process. They write what's obvious to them. And they often haven't identified the needs of their learners.

- **These same gurus may draft documents and then turn the drafts over to technical writers to edit, format, and polish.** Unless the technical writer has an opportunity to learn the subject matter intimately, many of the steps that may have been overlooked by the gurus aren't identified by the writer or editor. This process does, however, produce a document that may be more pleasing to the eye.
- **A technical writer is called in from the onset of a project.** The writer works with the developer who's the subject matter expert (SME). They work as a collaborative team, each adding their expertise to the project. This approach is the best of all possible worlds.

It's About Strategy, Not Software

Anyone who writes technical documents must understand how critical it is to take a strategic approach. For example, if you design a custom home, do you first call someone to wield a hammer? Of course not. A hammer is merely a tool. To design a custom home, you call an architect — a trained professional who designs layout; renders plans for the plumbing, electrical, and heating systems; and provides the structure. Then you call someone with a hammer.

The same holds true in technical writing. Effective technical documents require an information architect — *a technical writer*. Whether this person is a professional technical writer or an engineer or software developer who writes technical documents, this person must plan, design, and provide logical structure. Anyone can learn to use the software to create the document. Much like the hammer, software is merely a tool. The key to writing a great document is *strategy*, not software.

SHERYL SAYS

Someone once told me that they wouldn't make a good technical writer because they can't even use jumper cables to rev up an ailing car battery. Remember that technical writing isn't about jumper cables or about understanding every aspect of the technical and scientific communities. And it isn't about knowing every nuance of the latest software application. Very few people have that broad a knowledge base. Technical writing is about using *strategy and resources* to write clear, accurate, and logical documents. If you apply a logical strategy and avail yourself of resources, you can write just about anything — from using your instant pot to assembling a jet engine.

What You Need to Succeed

Following is a snapshot of what it takes to write clear and understandable technical documents:

» **Show respect for your learners.** You must always write with respect for your learners. Write with a positive attitude, not with arrogance. I've heard technical writers speak of their learners arrogantly with remarks such as, "We don't write these documents for idiots. If they're that stupid, they shouldn't be using this product." (Well, excuuuuse me.) Even the brightest humanoid may experience confusion when presented with something entirely new.

» **Pay keen attention to details.** You show your keen eye for detail in the way you think about what you write. I recently saw a resume from someone who was applying for a position as a technical writer. Check out these vague bulleted items I extracted from that person's resume:

- **"Contributing writer to city newspaper."** Which newspaper? What do they write about? Did they contribute weekly? Bimonthly? Monthly?
- **"Generated technical reports."** What do they mean by "generated"? Did they write the report or merely click the print button and spew it out?

This writer wasn't able to identify the details that a potential employer would want answered. Therefore, it was obvious that this person couldn't identify the details learners would need. *You must be able to anticipate the questions your learners will ask, and you must answer them.*

TECHNICAL WRITING BRIEF

» **Know your learners and their requirements.** Unlike a piece of communication you may write for a specific learner, you often write technical documents for a diverse group of learners. You must understand their needs in order to determine how you write the document and whether print or electronic media is appropriate, or whether streaming or simulations may work better. Check out Chapter 3 for information on understanding your learners by using the Technical Writing Brief.

CROSS REFERENCE

» **Collaborative efforts.** To tweak John Donne's famous quote: "No tech writer is an island." Even if you're the only tech writer on your project, you'll work with people inside or outside the organization: SMEs, editors, publication/electronic specialists, and end users. Chapters 2 and 14 talk about what it takes to work collaboratively.

» **Demonstrate the ability to leg-o your ego.** Last but not least, you must be able to leave your ego at the door. Your finished document will often be very different from your original draft. Everyone who reviews the document will feel compelled to pick up a pen and mark it up. It's all part of the process — the need to make a contribution. So, be prepared to have your work edited, re-edited, and perhaps ripped to shreds. (There's no chapter on the latter. Just smile a lot and mumble to yourself.)

Seeing Is Believing

As mentioned earlier in this chapter, we come in contact with technical documents in our daily lives. Some documents are well written; others leave us scratching our heads and wrinkling our brows. What's the difference?

CROSS REFERENCE

Following are *Before* and *After* cases that give you a chance to look at two different approaches for writing and presenting information. I took each *Before* case from an actual document where the writer buried the key information. Each *After* case shows an improved version where the key information jumps out. (Check out Chapter 5 to discover how to visually prepare technical documents that call the learner's attention to the key issues at a glance.)

Case 1

The United States Air Force tabulated the weight of 80 officers. Following are key questions learners may have:

- What's the median weight?
- What's the highest weight?
- What's the lowest weight?

Before document

The following chart is from the *ESD Process Improvement Guide* prepared by the Electronic System Division, Air Force Systems Command, Hanscom Air Force Base, Massachusetts. It displays the weights of the 80 officers in column format. The learner must search through all 80 numbers to answer the questions.

Weights of 80 Officers

206	180	139	163	159
155	180	165	149	127
159	171	141	190	159
153	181	180	137	161
115	156	173	165	191
159	110	179	145	144

Weights of 80 Officers

150	206	166	188	165
127	130	172	180	147
145	150	156	171	189
190	200	208	169	139
130	128	155	185	166
165	187	159	178	169
147	150	201	128	170
189	163	150	158	180
139	149	185	129	169
175	189	150	201	175

After

The distribution of the data in the form of a histogram shows the answers to the learners' questions at a glance.

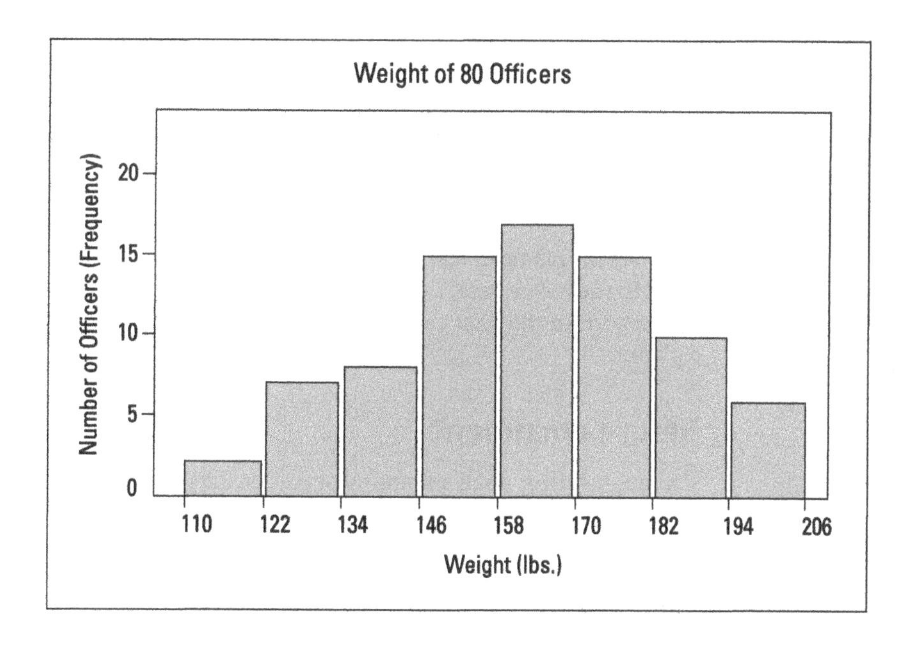

Case 2

The U.S. Department of Transportation wrote a "Transit Security Procedures Guide." Key issues in the following paragraph would be: What action must authorities take to prevent similar incidents? The *Before* paragraph buries this information; the *After* paragraph creates bullets to call attention to it.

Before document

> Training must include the sensitive handling of complaints and reports of offenses. Although these offenses are not actual assaults, there are victims. Authorized personnel must respond in a manner that leads the victim to believe that a similar incident will never happen again. This includes taking immediate action to apprehend the offender, taking information from the victim that may lead to arrest and prosecution, and notifying the police.

After document

> Training must include the sensitive handling of complaints and reports of offenses. Although these offenses are not actual assaults, there are victims. Authorized personnel must respond in a manner that leads the victim to believe that a similar incident will never happen again. This includes
>
> - Taking immediate action to apprehend the offender
> - Taking information from the victim that may lead to arrest and prosecution
> - Notifying the police

Case 3

Learners of this technical report want one question answered. What are the findings? In the *Before* case, the learner must read through an entire paragraph to find the answer. In the *After* case, the answer is part of the headline and identifiable at a glance.

Before document

The engineering analysis and observation did provide significant information. Review of the data indicates that a 10- and 12-mil printing process can produce consistent and acceptable result. The solderpaste heights were consistent and had a low standard deviation. The 8-mil spacing solderpaste heights were much lower with a larger standard deviation. Solder shorts and opens were much lower for 1- and 12-mil spacing as compared to the 8 mil. From a statistical point of view, the analysis indicates that no statistically valid conclusion may be drawn, due to the error in the model. Therefore, we must perform additional tests.

After document

Due to an error in the model, we must perform more tests.

The engineering analysis and observation did provide significant information. Review of the data indicates that a 10- and 12-mil printing process can produce consistent and acceptable result. The solderpaste heights were consistent and had a low standard deviation. The 8-mil spacing solderpaste heights were much lower with a larger standard deviation. Solder shorts and opens were much lower for 1- and 12-mil spacing as compared to the 8 mil. From a statistical point of view, the analysis indicates that no statistically valid conclusion may be drawn, due to the error in the model. Therefore, additional testing must be performed.

Beyond the Writing

Whether you're a newbie, certificate or degree holder, or seasoned technical writing professional — whether you freelance or work for a company — always have a current portfolio. It could be the decisive factor in whether you land your next assignment. Technical writing is project-based, and when one assignment is finished, you'll be on the hunt for another. The portfolio is your opportunity to showcase your range of writing talents, thought processes, tools in which you're adept, and other complementary skills.

Store your portfolio on your blog or on a portfolio site such as Contently or LinkedIn. Have a paper-based portfolio as well. Start creating your portfolio now and keep it up to date. Don't procrastinate.

Create a dynamic portfolio

If you're a newbie with no degree or certificate, search the Internet for "poorly written technical writing examples." You'll find loads. (Make sure to state they're not your documents, but represent your skills in dramatically improving documents.)

1. **Download three or four poorly written examples and label them "Before."**
2. **Read Chapter 5 about the elements of good technical writing.**

   This includes lots of whitespace; sentences limited to 20 syllables or fewer; and paragraphs limited to eight lines of text. Include strong, descriptive headlines, bulleted and numbered lists, and charts, graphs, and tables when applicable.
3. **Then recast the documents and label them "After."**

If you earned a degree or certificate, pull out five to six of your best works — those for which you received high grades. If papers show an A or A+ include the outstanding grades. Be prepared to discuss the following:

- The nature of the assignment
- Your strategy
- Tools you used
- Whether you were part of a team or worked independently
- Constructive criticism and why it was helpful

If you're a seasoned technical writer, include your best work for a variety of documents (reports, manuals, video scripts, and so on).

What if your writing is proprietary? Ask the person who engaged you if you can redact or change some of the information and showcase fragments. If not, recast the document. Change the names and any content that points to the original. Be prepared to discuss the following:

- Your writing process
- Whether you were part of a team or worked independently
- Whether you interviewed SMEs
- How long the project took
- Tools you used
- Did you finish on time? If not, why?

Every portfolio should include your bio or resume. In addition to the obvious (name and contact information), include the following: education, experience, technology tools, skills, foreign languages, and anything else that will put your best foot forward. Have a basic structure, but tailor each bio or resume to the project you're seeking.

You can also create an online portfolio and/or website to showcase examples of your writing. Platforms such as GoDaddy, Wix, and Squarespace have easy-to-use templates giving anyone the ability to put their work online.

Create a LinkedIn profile

LinkedIn is a free social networking site that focuses on professional networking and career development. Creating a profile will dramatically increase your network, position you as a professional, and let you reach out to high-quality learners and leaders. For a fee you can subscribe to LinkedIn Premium, which offers additional services such as classes, seminars, and who's searching your profile. (I used to have a robust website but cancelled it because most new business was coming from people finding me on LinkedIn.) Put yourself out there.

Present your business card

A business card succinctly conveys what your business is about. If you're a freelancer, it's your "calling card." With everyone having smartphones, electronic (digital or virtual) cards are rapidly replacing paper cards. It's best to use the advantages of both to leverage online and offline opportunities. Neither is expensive to create (some are even free) and they serve different purposes.

- **Paper cards offer a personal touch.** When you hand someone your card, you create an instant impression and build rapport. Receivers can jot down where they met you and if there's any follow-up (such as meeting for coffee). They're great for networking at events, exhibitions, and meetings.
- **Electronic cards are a quick way to share information online.** They can be created and shared via an app or produced as a PDF and shared on email, social media sites, and vCard files.

Regardless of which you use, take advantage of the back and front. On the front, have critical information: name, address (if relevant), phone, email, website and/or LinkedIn, and tag line (if you have one). On the back, share a testimonial, tag line, statistic, or relevant quote. As an added bonus, include a QR code to showcase your portfolio, testimonials, or any other brag pieces. It's easy to generate a QR code through the following vendors as well as others:

- **Printed cards:** VistaPrint, Zazzle, or Staples
- **Electronic cards:** Beaconstac, HiHello, Knowee, LinkTree, or QRD

RESOURCES FOR TECH WRITERS

Whether you're a seasoned professional or just starting out, avail yourself of these resources to move to the top of your profession:

- Google offers a free technical writing course that many have rated high for quality and a well-planned approach. It's geared for newbies and senior tech writers. Check out `https://developers.google.com/tech-writing`.
- Paying platforms for courses include Twilio Voice, Digital Ocean, Draft.dev, Vonage, Neptune.ai, and others. Fees range from $50 through $1,000.
- If you're a job seeker, check out `https://technicalwriterhq.com/career/technical-writer/technical-writer-interview-questions/` to get help with interview questions.
- Join the Technical Writer forum on LinkedIn.
- Sign up for a free course at `https://seofordevs.com/` to learn about search engine optimization (SEO), which will help you tailor content to the right audience while also driving traffic.
- Go to `www.isitwp.com/headline-analyzer` to analyze how your headlines are performing.
- Check out `www.writerswrite.com/technical/resources/` for additional resources.
- Visit freelance job sites such as `fiverr.com` and `upwork.com`.

Also, if you're a freelancer, join your local Chamber of Commerce and become part of your business community. Some of the benefits are business exposure, networking opportunities, mailing list access, seminars, freebies and discounts, members-only insurance, and more. If you work for a company, chances are your company belongs to the chamber, entitling you to attend meetings.

Tech Writing Career Trajectories

If you choose, technical writing can open doors to many fields and put you on track to an advanced career path. You could become a senior technical writer or project leader. If you're analytical, try your hand at writing for the medical or scientific professions or writing educational texts. You could become a technical journalist or online content developer, write scripts for technical videos, and more. With a technical writing background, your opportunities are vast. The following sections discuss a few career paths you can consider.

Translate technical documents

Companies with foreign suppliers or companies entering foreign markets are always on the lookout for technical translators. Documents that require translation include manuals, policy and procedure manuals, user guides, instructions, technical specs, scientific papers, patents, technical proposals, whitepapers, online help, and much more. There is a big need for technical translators in all fields. If you're fluent in a foreign language, this could be a niche for you!

While some companies use electronic translators, others prefer the human element. Electronic translators may have a low level of accuracy, and accuracy must be consistent across different languages. Machine-generated translations often sound robotic, choppy, and not culturally aligned. Machines can translate content, but they do make mistakes that can damage the brand that will be costly to a company. Although machine translations have come a long way, they're still far from ready to replace people.

Become a UX writer

The Internet is experiencing a digital renaissance as more and more companies realize that an excellent user experience leads to customer loyalty, which means greater profits. This has resulted in a burgeoning career path for technical writers — UX (user experience) writers. UX writers focus on the way users interact with and experience products, services, interfaces, and systems. They write microcopy with a an eye on clear messages of errors, notifications, calls to action, in-app tool tips and tutorials, product and service descriptions, UI button text, form fields, troubleshooting, and much more.

What does it take to be a UX writer? Understanding user research fundamentals and user-centered design, speaking the user's language, and empathy (aka *I've walked in your shoes.*) UX writers know that messages such as "An unknown error has occurred" or "404 not found" leave users dissatisfied and create a poor user experience.

Formal UX training isn't a must, but it can be helpful. There are courses for free and for a fee. If you're new to UX writing and want to get your feet wet, check out "free ux writing courses." If you've had some experience and want to refine your skill set or advance to the next level, you'll find fee-based courses at `https://careerfoundry.com/` as well as other sites.

Soar into the cloud

Have you ever thought of becoming certified in cloud computing to move your career in new and exciting direction? Cloud technology is one of the hottest buzzwords in technology, and businesses are clamoring for technical writers who are "certified" in the cloud. Check the Internet for "cloud certification" and you'll find a host of companies offering certification. This field creates an opportunity for tech writers to learn current authoring tools and applications to excel in this new technology.

By the way, have you ever wondered where the term *cloud* originated? The phrase cloud computing was inspired by the cloud symbol that's often used to represent the Internet in flowcharts and diagrams, as you see in Example 1-1.

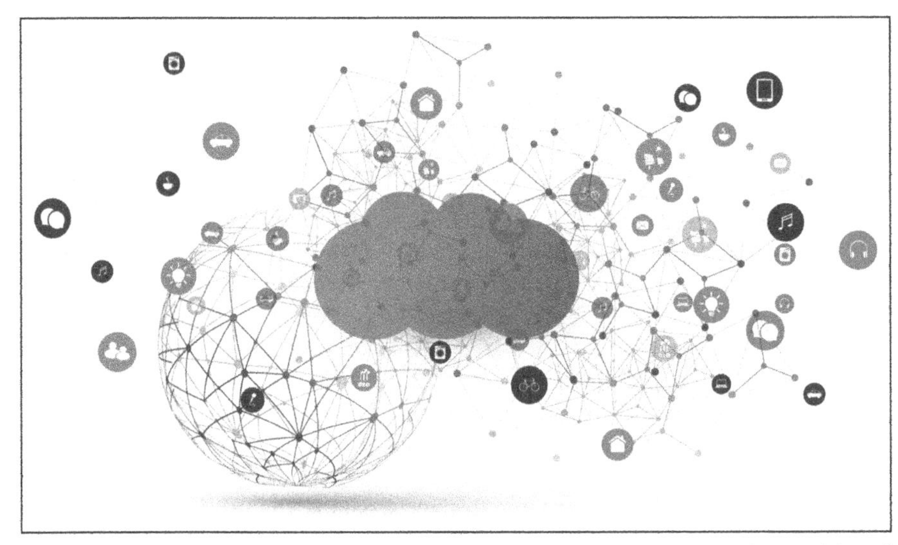

EXAMPLE 1-1: Clouds connecting computer hardware.

bagotaj/Adobe Stock

Become a scrum master

If you want to transition from a technical writer to a communications leader, consider becoming a scrum master. Scrum stands for *systematic customer resolution unraveling meeting*. It's a framework centered around the key principles of continuous improvement, flexibility, and respectful teamwork. Although the name sounds ghastly, it's basically a framework designed for small groups of ten or fewer. Groups break work into goals that can be completed within time-boxed iterations, called *sprints*. If you're familiar with rugby, the scrum encourages teams to organize while working through problems, much like technical writing teams need to do.

The skills of technical writers and scrum masters are aligned, so the transition can feel natural. Both professions require good communications skills, adaptability, plus information gathering and distribution savvy. As a scrum master, you'll manage teams of technical writers, UX writers, and project planners. You'll troubleshoot and be involved in stakeholder relations. If you're interested in becoming a scrum master, there are many online courses. Use these search words to find sites for each:

- Agile Scrum Master certification course
- Professional Scrum Master (PSM) course
- Agile Crash Course: Agile Project Management; Agile Delivery

MY JOURNEY AS A TECH WRITER

This is the tale of how I got started as a tech writer with no formal education in the discipline. Driven by goals, guts, and fortitude, I carved out a fulfilling and multi-faceted career. You can as well!

My Technical Rite of Passage

I graduated from college with a BA in Education and an MA in Business. After teaching English at the college level for several years, I heard the business world calling me. But how could my expertise in Chaucer and Donne answer that call? Through networking, I landed a tech writing contract with a chap who was starting a software company. I had no tech writing experience and he couldn't afford to pay much, so we were a matched pair. I did lots of research into the world of technical writing and produced several successful user manuals. To help his business grow, I launched a marketing campaign that got him into several trade shows and prestigious publications. His business took off. And so did I . . . elsewhere (my choice).

I fizzed with excitement and was ready for the big stage! I landed a contract assignment at Western Union (WU) that turned into a full-time position (back when WU was a full-fledged technology company). I was there for five years. Great experiences and great people. Then I got married and moved from New York to Boston, where I knew no one. (This predates working from home, so that wasn't an option.)

Onward and Upward

Through several WU relationships, I got a contract assignment with the U.S. Department of Transportation (DOT). I wrote several user manuals then was ready for another

(continued)

(continued)

challenge. Still at the DOT, I wangled my way into video productions. I had no experience but continued to be resourceful. I learned the art of storyboarding and wrote the script for several high-level videos on aviation and ground transportation. There was no stopping me now! When that contract ended, I started my own communications company: Sheryl Lindsell-Roberts & Associates. I've worked in a wide variety of industries and bring in graphic designers and video producers as needed. Always looking for the next challenge, I wondered, "What's next?"

Reaching for the Stars

New ideas were sparking in the depths of my imagination. I started writing books, but it was a long, rocky road over many years before success. Fast forward and I landed a contract for *Technical Writing for Dummies* — one of four Dummies books I authored (and 1 of 26 books I ultimately wrote).

With that knowledge in hand, I developed a technical writing workshop that I facilitate at large organizations. Not forgetting my passion for video productions, I prepared and facilitate a workshop titled "Stories and Storyboarding: The Key to Influential Presentations," which I hope to turn into my next Dummies book. Stay tuned!

IN THIS CHAPTER

» Kicking off with a Technical Writing Brief

» Benefiting from the team experience

» Getting your arms around the document

» Choosing the right type of delivery

» Completing a production schedule and outline

Chapter 2

Putting Together a Team and a Plan

If everyone is moving forward together, then success takes care of itself.

—HENRY FORD

In this fast-changing, market-driven, high-pressure world, roles and responsibilities constantly change. You may work with a team on one project and another team on the next project. The one certainty, however, is that you won't write technical documents in a vacuum; you'll be part of a collaborative effort. A collaborative effort doesn't necessarily mean a large team of people. Your team may be small and consist of a technical writer, UX writer, and a subject matter expert (SME) or a technical writer and a reviewer.

Large teams may also include scientists, engineers, software developers, systems analysts, marketing and sales managers, financial wizards, and the like. Or a team may include other writers, SMEs, editors, publication/electronic specialists, design specialists, and end users. Any combination of these people (or others) creates a collaborative team. When possible, the entire team should be part of the planning process.

Benefiting from the Team Experience

The face of teams has changed since the pandemic lockdown. Many people working from home were feeling socially isolated and have opted to return to the office. Others have opted to continue working at home. Still others chose a hybrid situation.

You'll often find yourself teamed up with diverse groups of people. Be prepared to incorporate their opinions, skills, and styles. Some people enjoy being part of a team; others don't. If you're one of the don'ts, try to view the experience as an opportunity to benefit from the wisdom and talents of others. You can't control the stock market, you can't control the weather, and you can't control those higher up the food chain. But you can control the way you work and play with others.

TECHNICAL WRITING BRIEF

Fill out the Technical Writing Brief when the entire team is assembled so you have everyone's buy-in. Chapter 3 discusses the brief in great detail and you can find a blank version in Appendix E and on the online cheat sheet (see the Introduction for instructions on accessing the cheat sheet online). If it's not possible to fill it out face to face, do it via collaborative tools or videoconferencing. Too often, the people involved in the project (even those responsible for signing off) don't get involved at the beginning. If everyone isn't on the same page from the start, the results can be time delays and missed deadlines.

Know who's on first

All team members need a clear understanding of their roles and the pecking order. Although this sounds basic, it's often the area where people trip over each other.

To simplify the process, prepare a "Who's Doing What Checklist" list to identify the tasks, leaders, responsibilities, and procedures. Feel free to pick from the checklist shown in Example 2-1. Once the team agrees to each one's responsibility, each member should sign the agreement and receive a copy.

CROSS REFERENCE

A significant risk to any project team is when one or more members miss learning critical information, especially during the planning phase. When teams are working remotely, collaboration tools help all members stay in contact throughout a project, thereby reducing the chances that someone could miss out on an important update or other information vital to the project. Check out Chapters 14 and 15 to learn more about how teams can benefit from collaborative document sharing and videoconferencing.

PUT YOUR EGO ASIDE

Two heads may be better than one, but two egos are worse. Typically, everyone who reviews a document feels compelled to comment. Doing so is just human nature. These comments aren't necessarily a reflection on your writing or your style; they're just part of the process. Never take others' feedback personally.

WHO'S DOING WHAT CHECKLIST

Team Responsibilities

- What specific tasks must be completed to finish the project?
- Who will be responsible for each task?

Working Procedures

- When will the team meet?
- Where will the team meet? (In person? Via video conference? As indicated for each meeting?)
- What procedures will be followed in the meetings?
- How will decisions be made (majority or consensus?)
- How will team members communicate? (Email? Phone? IM? Text? Other?)

EXAMPLE 2-1: Team checklist of who's doing what.

© John Wiley & Sons

Turn stumbling blocks into stepping stones

Viewpoints will vary, and conflicts will occur. These differences are a natural part of the collaborative process. Most conflicts, such as grammatical points, are minor. Others, such as basic approaches to a project, are more significant. View these issues as stepping stones, not stumbling blocks. Team members must find ways to work through all conflicts and make compromises. Doing so is all part of professional growth and helps you produce a top-notch deliverable.

REMEMBER

When you challenge a member's viewpoint, do it tactfully and offer a valid reason. For example, if you're writing a major report, you may say, "Your point is well taken, but do you think we should consider [whatever]?"

Choosing the Right Type of Delivery

Once you know the needs of your learners and have filled fill out the Technical Writing Brief, you'll need to consider these questions:

- **Do the learners have access to computers and the Internet?** For example, if you're writing a user manual for people who work on a shop floor, you need to know whether they have access to computers where and when they need information.
- **Will you need to update the information regularly?** For example, if you need to let your sales force know of on-the-spot product updates or price changes, you can do that electronically.

Table 2-1 shows the advantages and disadvantages of print documents. Table 2-2 shows the advantages and disadvantages of electronic documents.

TABLE 2-1 Traditional Print Documents

Advantages	Disadvantages
Everyone can read paper documents, even people who aren't very Internet savvy.	Make just one typographical error, and it's there for all the world to see until the next printing.
People are familiar with paper and some like its heft and feel.	It's expensive and time consuming to update documents. Commercial printing is labor intensive and costly.
You can pass documents around and share them with anyone.	You need physical storage space.

TABLE 2-2 Electronic Documents

Advantages	Disadvantages
Electronic documents are timely and interactive.	You lock out a small percentage of learners because not everyone has easy access to computers (although these folks are getting rarer).
You can provide links to other sites of interest to your learners.	There's always the risk of viruses, malware, and other "demons" even if you think the site is secure.
You can update documents on an as-needed basis.	When the computers are down, documents may not be available.
You can provide as many pages as you need because you're not limited by space.	Navigation may be a problem if the site isn't designed properly.

Completing a Production Schedule

If the team is to meet its deadlines, everyone must have a copy of the production schedule as early as possible. Team members don't always realize the impact of missing a target date, so you (or the team leader) may have to stress that point constantly. Although building in additional time is a good idea, it's helpful to include a column for both the target date and actual date. Then you can keep track of a slip in the schedule that endangers the delivery date.

Make the production schedule a "must"

Example 2-2 shows a production schedule (also known as a milestone chart) that you may want to use as is or amend for the needs of your project. Note that the target for completion is noted first (at the end). Then work backward. Even though the production schedule is listed in chronological order, once you identify the delivery date, you'll know how much time you have for each milestone and work toward it.

Timing is everything

The amount of time it takes to create a document varies among organizations and the people involved. For example, if key people aren't involved from the get-go, you may have to go back to the drawing board at a later stage in the process to incorporate their new viewpoints. Starting all over chews up a lot of time. To understand the process, analyze historical information on projects the organization has done in the past, and let past experiences be a measuring stick. Table 2-3 gives you some guidelines for timing issues.

You and others can work on certain phases of the project simultaneously to save time. For example, you can write the text while the graphics people prepare the artwork.

PRODUCTION SCHEDULE
[Name of Project]

Milestone	Target Date	Actual Date	Person Responsible
Fill Out Technical Brief			
Brainstorming Session			
First Draft Delivered			
Prepare Visuals			
Comments Due			
Final Draft Submitted			
Final Visuals Submitted			
Final Draft Approval			
Final Visuals Approved			
Mechanicals (if reqd.)			
Proofs			
Print			
Deliverable/Available	9/16		

EXAMPLE 2-2: Production schedule listing each milestone.

TABLE 2-3 Estimating Your Writing Time

Length of Document	Original Writing	Rewriting
Less than 100 pages	50 to 100 hours	25 to 60 hours
More than 100 pages	100 to 200 hours	60 to 100 hours

The Power of Brainstorming

Brainstorming is the process of moving ideas from your head. When the entire team holds a brainstorming session, it provides an opportunity for everyone to give and get input. This can be done face to face or via videoconferencing. Some-one needs to facilitate, and often the technical writer takes the lead. Although there are many ways to brainstorm, consider this visual example that works for lots of people:

- **Start by drawing a circle.** In the center of the circle, write the purpose of your project.
- **Draw branches and twigs extending from the circle.** Write your main ideas on the branches and your sub-ideas on the twigs.
- **Welcome input from everyone.** Don't pass judgment. This is only a brainstorming session, and any idea is worth capturing.
- **Don't dwell on any one idea.** The purpose of brainstorming is to get as much information as possible.

After you brainstorm, use the twigs and branches to create your outline, as shown in Example 2-3.

TIP

Look for existing documents or materials you can draw from. They may need to be updated but can still be valuable.

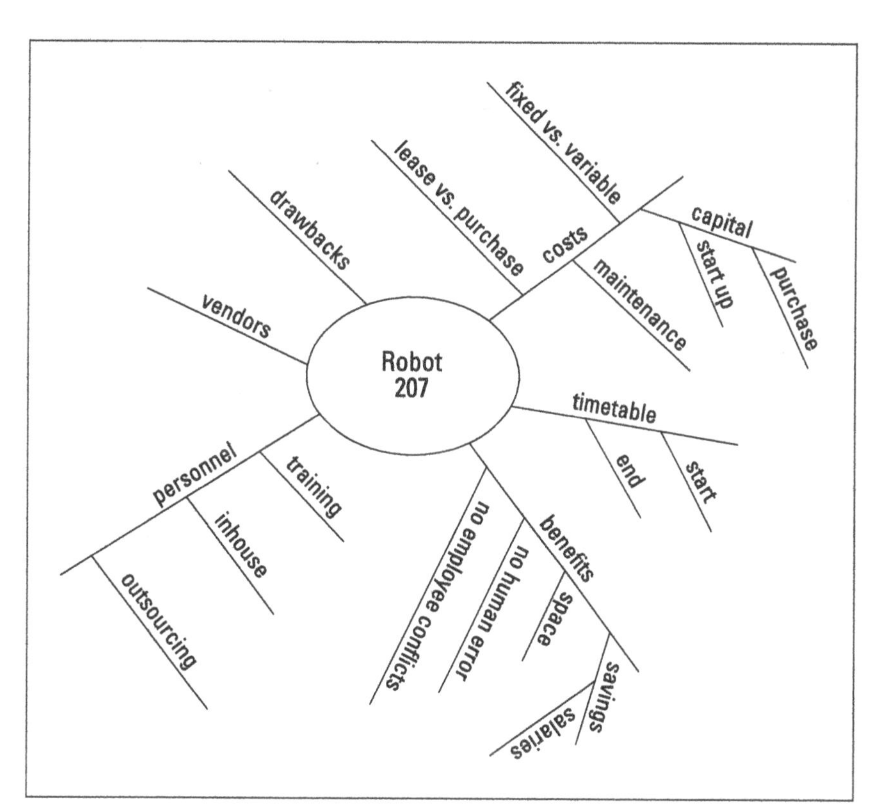

© John Wiley & Sons

EXAMPLE 2-3: Brainstorming on how to build a robot.

Generating an Outline

If time allows, generate the outline as a group activity, either face to face or online. Otherwise, this activity may fall on one person or a small team. When you complete the outline, get written approval from the key people on your project team.

Write a traditional outline

Outlining is a tried and tested method learned in high school. Some people love it; others hate it. Your computer software's outlining feature makes it easy for you to experiment with the organization and scope of information. Example 2-4 is a standard outline.

I. Create Message

- A. Preparing a Message
  - 1. Accessing the Create Message Menu
    - a. Introduction
    - b. Prerequisite
    - c. Procedure
    - d. Setting the Delivery Default
  - 2. Functions of the Create Message Screen
    - a. Forwarding a Message
    - b. Changing the Message Type
    - c. Verifying the Address
- B. Address Defaults
  - 1. Accessing the Address Defaults Screen
    - a. Introduction
    - b. Prerequisite
    - c. Procedure
  - 2. Functions of the Address Defaults Screen
    - a. Forwarding a Message
    - b. Changing the Message Type
    - c. Verifying the Address

II. Online Communication

- A. Communicating with IMS
  - 1. (and so forth)

EXAMPLE 2-4: Traditional outline for a user manual.

Use a decimal numbering system

The decimal numbering system outline shown in Example 2-5 is helpful for documents in which you may need to reference certain sections or paragraphs. If you use this system, you can say to a group, "Look at No. 2.1.2." They'll know exactly where to find the information.

Create an annotated table of contents

Another outlining option is to create an annotated table of contents, such as the one you see in the front of this book that breaks out the sections in the chapter.

1.0 **BACKGROUND**
- 1.1 PURPOSE
- 1.2 APPROACH
  - 1.2.1 Overview
  - 1.2.2 Tasks
  - 1.2.3 Critical Concerns

2.0 **TECHNICAL CONCERNS**
- 2.1 SUCCESS FACTORS
  - 2.1.1 Locating Expertise
  - 2.1.2 Developing Training Procedures
- 2.2
- (AND SO FORTH)

EXAMPLE 2-5: Using the decimal numbering system outline.

Getting Your Arms around the Document

The purpose of writing a technical document is to explain or report on a technical or complex subject. Therefore, unless you're the technical guru writing about something you know intimately, you must research the subject. Gathering data — learning all you can about the product or service — is the lifeblood of the project.

Conduct internal research

If your research is internal, garner your information by interviewing SMEs and others involved in the project, observing people in and around the workplace, reading spec sheets and company literature, and just snooping around. Here are some tips for keeping your nose to the grindstone without getting it snipped off:

» **Get to know the product or service intimately.**
  - Test it. Use it. Embrace it. Learn all you can.
  - Study system specs and engineering drawings.
  - Interview the SME. Bring a list of questions and don't be satisfied until all your questions have been answered.

» **Absorb the big picture.**
  - Find out about the purpose of the product or service, who'll use it, why, and when.
  - Learn the industry's jargon and acronyms.
  - Become familiar with the good and bad features.
  - Learn about any possible flaws, such as what features may break, malfunction, or cause problems.
  - Find out if there are any dangers or hazards you must make users aware of.

» **Cozy up to the marketing and sales departments.**
  - Learn the advantages and disadvantages of the product or service.
  - Understand the features and benefits you need to highlight.
  - Know how the deliverable will be distributed and updated.
  - Find out whether there are related publications, manuals, FAQs, or any other supporting data.

» **Find out whether you need to prepare unpacking instructions or a parts list.**
  - Find out what's in the shipping carton.
  - Learn whether users need instructions to install or assemble the item.

» **Gather as much written or graphical data as you can. You can always discard what you don't need.**
  - Decide what photos or graphics you need.
  - Get copies of a style manual or standard text that similar projects used.

TIP

It's important that you don't merely trust your memory. Take copious notes or talk into your smartphone (or some other electronic device) while doing your research.

Conduct external research

You do external research by interviewing people outside your organization or surfing the Internet. Here's some advice to make your job easier:

- **Interview people outside your company.** You can either visit people personally or send an email. Which to do, of course, depends on the nature of your research. If you need to gather information from a lot of people, you may find a questionnaire is appropriate. Check out Chapter 11 for tips on generating a questionnaire that evokes responses. If you need to meet face to face or via videoconferencing with one or more people, follow these guidelines:
  - Prepare a list of objectives and questions beforehand.
  - Let the interviewee do most of the talking. After all, you're there to learn.
  - Grab all the literature the interviewee is willing to share.
  - Take copious notes or record the session, with permission of course.
- **Surf the Internet.** If you don't know the URL of the site you want, access one of the popular search engines. Check out Chapter 17 for in-depth tips on using the Internet for research.

MAKING THE MOST OF PEOPLE'S TIME

According to Nancy Settle-Murphy, virtual leadership coach and facilitator and President of Guided Insights: "Many teams are being overwhelmed by excessive emails or a steady string of Slack notifications. Teams need to agree on how email, IMs, team portals, texts, and so on, are best used and under what circumstances. Agreeing on the time of communications is also important, especially when team members span multiple time zones. Is it okay to send emails or IMs during someone else's off-hours? Is it okay to wait to reply during your own work hours?" All those considerations should be agreed upon during the planning process.

2

The Write Stuff

IN THIS PART . . .

Get to know your learners, key in on what they need to know, pinpoint the most important issues, and understand what questions they need answered.

Learn to write a draft and integrate the revision process.

Harness the power of headlines and subheads, use bulleted and numbered lists, and include charts and tables when appropriate.

Keep it short and simple, use the active voice and positive words, be clear and succinct, and use inclusive and culturally sensitive terms.

Proofread carefully, edit for clarity and flow, and use readability assessments.

IN THIS CHAPTER

» Feeding a Martian

» Getting to know your learners

» Slicing and dicing the Technical Writing Brief

Chapter 3
Completing a Technical Writing Brief

I can't write without a learner – it's like a kiss, you can't do it alone.

—JOHN CLEEVER, AUTHOR

The key to writing a technical document is to have an in-depth understanding of your learners. You must understand who they are, the level of detail they need, how they process information, and how they'll use the information. If you don't gather that information, your document will be as ineffective as foreign language instructions that come with do-it-yourself (DYI) projects. The Technical Writing Brief is your key to unlocking this knowledge to creating usable, valuable content.

How to Feed a Martian

This is an exercise I use when I present technical writing workshops. At first, participants think I'm outrageous, even wacky (and you may too), but it proves a critical point. Please stay with me here.

Imagine this scenario: It's the year 2500, and business is conducted intergalactically. You have a hot deal pending, and it's hinging on the arrival of a business associate named Zeb from the planet Zeblonia. This is Zeb's first trip to the planet Earth. If you can razzle-dazzle him and get him to sign the deal, you'll get the big promotion you've been coveting.

You plan to be at the space pad where Zeb's spaceship will land. And you plan to welcome him personally and bring him directly to your office for lunch, followed by the all-important meeting. However, Zeb's ship is delayed because of heavy intergalactic traffic, and you can't be there because a crisis demands your immediate attention. You arrange for a driver to pick Zeb up and drop him off at your house. This will give Zeb time to eat and freshen up before he meets with you and the other Earthlings.

You realize that Zeb will probably be starved after his long and arduous journey; after all, he missed lunch. You didn't have time to shop (you weren't expecting him at your home), but you do want to make Zeb feel at home. The fixings for a peanut butter and jelly sandwich are all you have in the cupboard. Although it's not a gourmet meal, it should hold him over until later. You know that Zeb speaks a little English, so you leave him directions to make a peanut butter sandwich.

SHERYL SAYS

Your assignment, should you choose to accept it, is to write detailed instructions for Zeb on how to make a peanut butter and jelly sandwich. *I hope you'll try this exercise because there's a method to my madness.*

How to make a peanut butter and jelly sandwich.

____

____

____

____

____

____

Will Zeb go hungry?

What did you learn from this exercise? Perhaps nothing, but I hope that isn't the case. Check out some questions you need to think about in order to write clear instructions for Zeb to have a shot at understanding you:

» What's Zeb's level of understanding?

- Although he speaks English, would he understand your terminology?
- Would Zeb know a jar of peanut butter from a jar of jelly?
- Would he even know what a jar is?

» Did you give Zeb all the information (and steps) he needs?

- Would he understand how to remove the lid from the jar?
- Would he even know that he should remove the lid?
- Is your peanut butter from a health food store? If so, did you tell Zeb to mix the oil with the gooey stuff?

» Did you present the information for quick understanding?

- Will Zeb understand your terminology?
- Would visuals help? (Perhaps you could draw a picture of the peanut butter jar.)
- Would a combination of words and graphics be appropriate?

» Did you achieve your purpose?

- Was Zeb able to make a peanut butter and jelly sandwich?
- Was he smiling?
- Was he no longer hungry?

Fill Zeb's empty stomach

Most participants start out thinking that this exercise is quite easy. They begin showing PowerPoint slides with basic directions much like these:

1. **Open the jar of peanut butter.**
2. **Open the jar of jelly.**
3. **Smear the peanut butter and jelly on the bread.**

Okay, these instructions may be simple for the average Earthling, but would Zeb know a jelly jar from a loaf of bread? In the scenario, I said that Zeb speaks English, but do you understand everything you read just because you speak English? For example, would you be able to distinguish *montmorillonite* from another mineral alongside it? Do you even know what the word means? Just because you speak English doesn't mean you understand everything you see in writing. You understand only what you've been exposed to.

Of course, you won't be writing technical documents for aliens from another planet, but much of what you write may be alien to learners on this planet. The bottom line: It's vital that you understand your learners so they'll understand you.

Getting Jump-Started with the Technical Writing Brief

Before you write a technical document, realize how and why people read these documents. They don't read them as they read engaging novels. They don't put their feet up on the table in front of a roaring fire and bury themselves in every word. And they don't read them for pleasure.

People generally read technical documents as references, often to figure out what they did wrong. In that case, they're frustrated and don't want to pore over gobbledygook. (Chapter 6 is chock-full of nifty tips for cutting gobbledygook.)

As an experienced technical writer, I never commit one word to my computer until I've completed a Technical Writing Brief. It's such a critical part of the writing process that you'll find it in several places:

- Example 3-1 in this chapter, with a full explanation.
- Appendix E, so you can flip to it for reference.
- Online at `www.dummies.com` (search for Technical Writing for Dummies Cheat Sheet). I suggest that you print this out and keep it where it's handy.

You'll also notice icons placed throughout this book as a reminder to complete the brief before you start writing. Feel free to use it as it appears or amend it to suit your project. After the first time, it'll save you hours.

After working with this brief for a short time, many learners report that, by completing it before each writing project, they cut writing time by 30 to 50 percent. That's huge!

Slicing and Dicing the Technical Writing Brief

TECHNICAL WRITING BRIEF

Example 3-1 shows the Technical Writing Brief. As you go through the chapters, you'll see the Technical Writing Brief icon as a reminder to complete it before starting any document. After you fill it out once, it will be easy from then on.

REMEMBER

Your purpose at this stage is to plan your document, not begin writing it.

The following sections explain each section of the Technical Writing Brief in detail.

About the document

1. **Type of document:** What type of technical document is this? User manual, product description, reference manual, whitepaper, abstract, spec sheet, presentation, article, report, eLearning document, or something else? Will it be presented as a paper-based document or be online? Will it be a video (streaming)? Or will it be a simulated learning experience (SLE)?
2. **Presentation context:** How will the document be presented? On paper, online, via streaming, in a simulation, or some combination of these? If this is part of a project that has other components (such as training, supporting literature, spec sheet, or others), indicate what they are.
   - Is this a standalone writing project that has no other components?
   - If this is part of a larger project, can you coordinate efforts with another team?
   - Are there aspects of either project that can be done in tandem?
   - Can you share any information?
   - Have design or style issues been identified?
3. **Target date for completion:** What is the real drop-dead date? (Too often people pick an arbitrary date, and then everyone gives up eating and sleeping to meet it. They later find out the deadline wasn't a real one.) Once you determine the date the document is due, work backward to fill out the production schedule found in Chapter 2.

About the Document

1. Type of document
2. Presentation context (Paper? Online? Streaming? Simulation? Combination?)
3. Target date for completion

Learner Profile

4. Who are the learners?
   - A. Are they technical, nontechnical, or a combination?
   - B. Are they internal (to your company), external, or both?
   - C. Do you have multi-level learners?
   - D. If so, what percentage are there of each?

5. What do the learners *need* to know about the topic?
   - A. What's their level of the subject knowledge, if any?
   - B. How do they process information?
   - C. What jobs do they perform?

6. What is their attitude toward the topic? (Positive? Neutral? Negative?)

Key Issues

7. What are the key issues to convey? (A is the most important.)

   A. ______________________________
   B. ______________________________
   C. ______________________________
   D. ______________________________
   E. ______________________________

Budget

8. ______________________________________

Project Team

9. Who's who on the project team?

   ____
   ____
   ____
   ____
   ____

EXAMPLE 3-1: Technical Writing Brief.

Milestones

10. List the milestones plus anticipated dates.

Milestone Date

____

____

____

____

____

Approval Cycle

11. What's the approval cycle? (Start from the bottom, with A being final reviewer.)

A. ______________________________________

B. ______________________________________

C. ______________________________________

D. ______________________________________

E. ______________________________________

EXAMPLE 3-1: Technical Writing Brief.

Learner profile

4. Who are the learners? Identify your relationship with your learners. Do you share similar experiences and educational backgrounds? Are they familiar with your product or industry?

For example, a small detail such as knowing their approximate ages may be important in deciding the output. Surveys show that younger people are more apt to use online documentation than older folks. (Some of you may not be too delighted to hear that *older* means over 40.) Another may be that younger people grew up in the computer age and relate to gamification, streaming, and/or simulations.

A. Are they technical, nontechnical, or a combination? This will help to determine the backgrounds of your learners so that you can use appropriate language and references. For example, do they share a common background with you or with each other?

B. **Are they internal (to your company), external, or both?** For internal documents, identify your learners by name and job function. For external documents, identify categories of learners (managers, engineers, and others).

C. **Do you have multi-level learners?** If you're writing to multiple-level learners, rank them in order of seniority. For example, if your learners are a mixed audience of managers, techies, and salespeople, consider dealing with each group separately in clearly identified sections of your document. (Also, determine any language problems.) Here's how you may want to structure the elements in a report for multiple learners:

- **Table of contents:** This creates a pathway for everyone. (Notice how the *Dummies* books have an abbreviated TOC and an expanded one.)
- **Executive summary (or abstract):** Designed for the managerial level — those who want the big picture only. Find out more about abstracts in Chapter 9 and executive summaries in Chapter 13.
- **Appendix:** Appeals to the techies — those who want all the nitty-gritty details, including data tables.

D. **If so, what percentage are there of each?** This information is critical to help you structure the document.

5. **What do the learners *need* to know about the topic?**

A. **What's their level of the subject knowledge, if any?** Think of what your learners *need to know* — not what they already know. You don't want to give too much or too little information. For example, if you're writing documentation for a software application, you must know the learners' level of computer skills. Following are a few examples:

Greenhorns may have limited knowledge. They're prime candidates for the heft and feel of the printed page, not the electronic page.

Sporadic users may have used the system, but they don't use it often enough to remember the commands and other good stuff. They may be amenable to online documentation, if it's easy to use. Or a combination of paper-based and electronic documents may be appropriate.

Aces are the true power users. They understand the ins and outs of the product but may have occasional questions. Their manuals may prop up their screens, and they're prime candidates for online documentation, streaming, or even simulations.

- What's their level of knowledge about the subject?
- What acronyms, initials, or abbreviations will you need to explain?
- Do they have any preconceived ideas?

- What are the barriers to their understanding?
- Is there anything about their style of dealing with situations that should drive your tone or content? Chapter 6 has a full discussion about using the proper tone.

B. How do they process information? During my years of experience in the field, I discovered that people with academic, scientific, or technical backgrounds tend to be process oriented. They benefit from step-by-step explanations. Those with backgrounds in business or law are answer oriented. They respond to quick answers. Creative types are usually visually oriented and benefit from charts, tables, and any visual representation. Discover more about preparing visuals in Chapter 5.

C. What jobs do they perform? Are your learners CEOs, managers, engineers, administrative assistants, data entry specialists, or shop-floor personnel? For example, people on a shop floor need hard-copy instructions because they wouldn't necessarily have ready access to a computer. Managers are big-picture people; they want to know the key issues. Technical folks want the details. Salespeople need to know the benefits.

6. **What is their attitude toward the topic? (Positive? Neutral? Negative?)** You may not always tell your learners what they want to hear, but you must always tell them what they need to know. Your learners' attitudes fit into one of these three categories. (You also see an example of what may invoke the reaction.)
   - **Positive:** You anticipate that the project will be completed one month early. (Who wouldn't be happy to hear that?)
   - **Neutral:** You suggest that everyone stay the course. (Neither good news nor bad news.)
   - **Negative:** You recommend to management that they hire ten more engineers. (That means issuing more paychecks, thereby cutting into the bottom line.)

Key issues

7. **What are the key issues to convey?** Every document has a purpose and key issues to convey. If your learners forget everything else, what's the one key issue you want them to remember? (This is akin to putting on an advertising cap to prepare a ten-second spot commercial.) Then proceed to the second most important and so on.

Budget

8. **If the writing team has any budgetary responsibilities, this is where that information goes.** If not, leave this blank or put N/A for not applicable.

Project team

9. **Who's who on the project team?** List all the people involved in the project: subject matter experts (SMEs), writers, editors, graphics folks, videographers, and any other "worker bees."

Milestones

10. **Along the way there will be several milestones to make sure the project is on target and on budget.** Note each milestone and the date it will be determined. If you're not meeting your milestones, you may need to reconvene the team.

Approval cycle

11. **Consider everyone who'll provide input and mark up your pearly words of wisdom.** Don't be discouraged when this happens. If people don't put marks on the page, they don't feel that they did their jobs. It's nothing personal.

    The final reviewer will probably have minor changes because the document will have been through many review cycles.

GETTING TO KNOW YOUR LEARNERS

Gathering information about your learners doesn't have to be a daunting task. More and more industries collect speculative information about learners when they plan a product. If you don't have that information, following are ways to gather it.

Before You Write

Try to get as much information as you can to reflect the needs of your intended learners and the problems they may have. Follow these tips:

- Get some test subjects (guinea pigs) and give them a real-life assignment. Don't give the testers any documentation. Instead, provide a subject matter expert (SME). Have the SME make notes of the questions the testers ask and what they can figure out on their own.

- Do a needs analysis by sending a survey to a broad sampling of intended users.
- Hold training classes before an alpha test. Listen to the questions the students ask and how frequently they're asked.
- Interview intended users. For example, if you're documenting software for the classroom, talk to teachers or observe them in the classroom.
- Look at what the competition is doing.

After You Write

You also need to collect some information after your project is completed. Doing so can help you with rewrites of your document. Follow these tips:

- Get feedback from alpha and beta sites.
- Send an evaluation form to get the scoop on what users really think. Check out Chapter 11 for tips on writing questionnaires that get results.
- Get input from the folks on the help lines. Compile a list of the 25 most frequently asked questions. Check out Chapter 8 for information about writing FAQs.

IN THIS CHAPTER

» Psyching yourself up

» Writing your draft

» Editing and revising your draft

Chapter 4
Crafting a Draft

Get down to it. Take chances. It may be bad, but it's the only way you can do anything really good.

—WILLIAM FAULKNER, AUTHOR

You may be wondering why this chapter is so short. When you use the Technical Writing Brief (discussed in Chapter 3) and you plan properly (as Chapter 2 explains), you've provided the structure to draft your document with ease. With proper planning, writing the draft is just like adding meat to the bones. If you don't plan properly, you may sit and sit and stare at your computer hoping that pearly words appear on the screen. They won't.

Psyching Yourself Up

The most difficult part of writing the draft is getting started. Therefore, don't wait for inspiration; just sit down and do it. Following are some helpful hints to put you in the "write" frame of mind:

- **Get comfy.** Try to get as comfortable as you can because comfort enhances concentration. For those who work in cubicles or shared office space, creating the environment you want isn't always easy, but you can do a few things for yourself. If you like to air your feet, slip your shoes off. If you like to munch, get a bag of snacks. If you like music, put on a headset.

- **Gather what you need.** Gather all your notes and reference materials before you start. When you constantly get up to find what you need, you break your train of thought.
- **Try freewriting.** It doesn't matter what you write. It could be a shopping list. The idea is to relax your mind and get something down on paper. Set a time for yourself, even if it's just ten minutes, but keep writing. Many people find that once their minds are relaxed, their attention turns to the draft and they can get started.

Getting Down to Business

This is where all your planning pays off. You already have a direction (in whatever form you generated it); therefore, writing the draft is akin to filling in the blanks. Here's how to proceed:

- **Write one section at a time.** Start with the section that's the easiest to write. Your learners will never know where you started.
- **Avoid the temptation to go over what you write.** The important thing is to keep moving forward.
- **Keep moving forward.** If you can't think of the right word, use another and keep going. This isn't a finished document — it's a draft.
- **Set reasonable timeframes.** Once you're in the zone, set a reasonable period of time for each segment of your writing. You may set a goal to write for a half hour or an hour. Continue to write until you meet your goal even if the words aren't flowing. If you're on a hot streak, by all means keep chugging along.

SHERYL SAYS

Don't get it right. Get it written. It's a draft — your first offer. Don't worry at this point about spelling, grammar, or punctuation. Save that for the proofreading and editing phases, when you'll trim the fat and end up with a gourmet repast!

After you finish each writing session, get some distance. Even if you're on a tight deadline, take a five- or ten-minute break. Put your feet up. Go for a brief walk. Return an email. Get a cup of coffee. Pat yourself on the back. You need to clear your head and refresh your brain.

CROSS REFERENCE

After you write the draft, incorporate the guidelines in Chapters 5, 6, and 7 before you send your document out for review.

- **Chapter 5 talks about how to make your documents visually appealing.** Your documents need strong visual impact in order for people to want to read them.
- **Chapter 6 discusses honing the tone — how you "sound" to your learners.** Your words tell your learners a lot about you because they create a window to your mind. You always want to "talk" to your learners in a tone that's appropriate for them. For example, use *you* and *your* rather than "the learner should. . ."
- **Chapter 7 explains ways to proofread and edit effectively and assess the readability of your document.** It also has a great checklist to use for all your technical writings.

Integrating the Editing Process

Rarely will you write a document without getting input from others — before, during, and after the writing process. In a long technical document, you may have several rounds of revisions.

The wallpaper edit

Here's a nifty process that works well when the team is assembled in one place. Unlike using collaboration tools (more about that in Chapter 14) wall-paper editing enables reviewers to see the continuity of the pages, as you see in Example 4-1. (Wallpaper editing isn't an actual term, so you won't find it in a search. It's a term I made up for lack of a better one. (Perhaps you'll find it in the OED one day.) Regardless of what you call it, here's how the process works:

1. **Print your document, double-spaced, and tape each page to a wall in sequential order using easy-release tape**. You may do this one chapter at a time or whatever works for the length of your document. You may also do this for each round of revisions.
2. **Schedule a date and time for each person or group to edit the document.** Provide a different color marker so you know who wrote what.

This process works well because it makes people commit to a time for editing or they lose their right to comment. (This also prevents the all-too-common reviewer who ignores the document until "someday," which isn't one of the days of the week.)

All you have to do is post the document on a wall, generate a schedule for people to edit, and make sure they know the time and place. Build this process into the production schedule and people know you mean business. Check out Chapter 2 for more information about creating a production schedule.

EXAMPLE 4-1: Wallpaper editing in real time.

Gorodenkoff/Adobe Stock

Hold on to your ego

Everyone who reviews the document will undoubtedly make at least one change. When people pick up a pen or edit on the computer, they feel compelled to change something. *Don't take the changes personally;* they're the nature of the beast. Ask the members of the team to review and edit the draft carefully and critically. Remind them to check for large and small issues — from the organization of each segment to technical accuracy. If the edits are extensive or differ dramatically from what the team planned, give team members the option of accepting or rejecting the changes.

Major changes may lead to another round of planning. That's not a negative. It's just another opportunity to evaluate the document and make it better. After all, until the time a document gets posted or printed, it's a work in progress.

Revising Your Work

Once you have all the edits, decide which changes to incorporate into the document. This time the edits, if any, should be minor and easy to include. You must know when to stop reviewing, however. At some point, you're not making the document better, you're just making it different. Stop when you notice any of the following:

- **You start nitpicking about insignificant words.** For example, you get wrapped around the axle trying to decide whether you should use *easy* or *simple.* Pick one and get on with your life.
- **You revise the revisions of the revisions of the revisions.** At some point, you start making changes for the sake of making changes. If you're not adding value, give it up.
- **You can't stand to look at the document one more time.** That's the telltale sign to pack it in.

IN THIS CHAPTER

» Using whitespace

» Sticking to sentence and paragraph length

» Harnessing the power of headlines

» Understanding the value of search engine optimization (SEO) in headlines

» Punctuating bulleted and numbered lists

» Realizing the power of visuals (graphs, charts, tables, and screenshots)

Chapter 5
Designing Documents to Enhance the User Experience

Visual design is the emotional bridge between functionality and the human experience.

—KIM CULLEN, DESIGN DIRECTOR

We're living in a digital world where information bombards us at every turn. Our brains can't handle the abundance without feeling overwhelmed. This is what makes visuals rise above the clutter. Research has taught us that about 65 percent of the population are visual (spatial) learners, meaning they need to see information in order to remember it and apply it.

This also relates to sentence and paragraph lengths, headlines and subheads, bulleted and numbered lists, word choice, and visuals.

Given that most of the documents you write will ultimately live in the digital world, a high readability index and a concern for user experience can be factors in search engine optimization (SEO) rankings. If people can't find what you wrote, they can't benefit from your knowledge. Impactful visuals not only play a factor in elevating SEO rankings, but they also grab and hold a learner's attention, giving key information at a glance.

You may have spent days, weeks, or months gathering information and writing a great document. If the document doesn't have visual appeal, however, nobody will read it or understand it. You don't need to be a graphic artist with fancy software to create a pleasing visual design that has impact on your learners. This chapter walks you through simple ways to turn *ho-hum* documents into *smashing* ones.

Grabbing Your Learners' Attention

Visuals serve as attention-getters to communicate information at a glance. They provide a subtle, unconscious signal that the document is worth learners' attention. When a document has visual impact, it attracts attention, invites learnership, and establishes the credibility of your message even before you state your case. Here's why:

- **Visual impact organizes information.** A good visual design breaks the document into manageable, bite-sized chunks, making it easy for learners to find the key pieces of information. Learners can concentrate on one idea at a time.
- **Visual impact emphasizes what's important.** You can create a hierarchy of information so your learners can separate major points from supporting ones — much like you see in newspapers or online news apps. In today's harried world where people are pulling their hair out because of tight schedules, your learners will appreciate a quick read.

Using Whitespace

Whitespace (or white space) is a key ingredient in visual design; it includes all areas on the page or screen where there's neither type nor graphics. (On a computer screen, this area is referred to as *quiet space* or *blank space*.) Without whitespace, our brains become overwhelmed and stop taking in information. Here's what whitespace does for your document:

- Makes it inviting and approachable
- Provides contrast and a resting place for the learners' eyes
- Creates the impression that the document is easy to read

Following are tips for using whitespace effectively:

- For paper documents, use 1- to 1 ½-inch top, bottom, and side margins to create a visual frame around all the text and graphics.
- For electronic documents, leave a ¼ to ½ inch margin all around.
- Double-space between paragraphs to help the learner see each paragraph as a separate unit. (Don't indent paragraphs. That convention dates back to the days of manual typewriters.)
- Emphasize key pieces of text (words, phrases, or paragraphs) with whitespace or a different font.

Giving Learners a Break

It's crucial to break your sentences and paragraphs into manageable, bite-sized chunks of information. Many technical writers use long sentences and dense paragraphs. Doing so makes technical information difficult to digest and causes learners to tune out. When you optimize sentence and paragraph length, you give your documents more visual appeal.

Limit sentences to 20 syllables

Of course you don't want to count syllables, and you don't have to. Turn on the reading index in your word processor and it will do the counting for you. The index gives you sentence and word information and highlights the "offenders." Your readout may look something like this:

Sentences > 25 syllables = 25 percent of your writing

Words > 12 letters = 2 percent of your writing

UP, UP, AND OY VEY

In its day, the Boeing 747 was considered an engineering phenomenon. It holds up to 490 passengers and is 2,775 inches long (that's longer than two basketball courts). This 710,000-pound marvel can leap tall buildings in a single bound. The 747 took five years to develop, which included hundreds of thousands of labor hours. Imagine what the documentation was like.

The documentation contained 31,000,000 (yes, that's 31 million!) sheets of instructions, plans, specifications, diagrams, parts, and change orders that included buzzwords, symbols, figures, and more. If every Boeing 747 tried to carry that tome on board, I wonder whether the plane could have taken off. Better yet, imagine keeping those tomes on your bookshelf and poring through them.

Vary the length of your sentences so you don't sound robotic. A great rule of thumb is to have a variety of longer and shorter sentences. The following example shows how a lengthy sentence of 39 words can be broken into three readable sentences.

- **Lengthy (39 words):** "As you see in Diagram A, variations across the die arise from stencil aperture dimension variations and stencil cleanliness, and smaller variations arise from random defects such as inclusions in the paste and contamination from the wafer or environment."
- **Just right (broken into three sentences):** "As you see in Diagram A, variations across the die arise from stencil aperture dimension variations and stencil cleanliness. Smaller variations arise from random defects. These may include inclusions in the paste and contamination from the wafer or environment."

Limit paragraphs to eight lines

Limit paragraphs to about eight lines of text, which is about five to six sentences (approximately 150 words). That's considered readable information and is best for SEO rankings. Think of your paragraphs as trains of thought. When one train leaves the station, another train arrives that heads in the same general direction. Although there are no hard and fast rules about paragraph length, when you limit each paragraph to eight lines, you have a very readable document.

EXPRESS NUMBERS IN WRITING

Experts don't always agree on how to express numbers in writing. Some say that one-word numbers should be written out, and two-word numbers should be expressed in figures. Others say spell out single-digit numbers from zero to nine, and use words for the rest. *The Chicago Manual of Style* recommends spelling out the numbers zero through one hundred and using figures thereafter — except for whole number. My rule of thumb is to use good judgment, and be consistent. Use numbers when you want to emphasize that number, and words when you don't. One rule everyone agrees is to never start a sentence with a number.

Correct: ***Three hundred people attended. . .***

Incorrect: ***300 people attended. . .***

- **Dense paragraphs:** When you write dense paragraphs, learners may find your text intimidating. They'll fail to see your subdivision of thoughts, and may even skip over those paragraphs.
- **Short, choppy paragraphs:** When you present learners with a lot of short, choppy paragraphs, it's difficult for them to see the logical relationship between your ideas and thoughts.

Harnessing the Power of Headlines

Newspapers, online news apps, and magazines use informative headlines as guideposts for visual impact. The headlines tell a story and direct people to what's important. When you write compelling headlines, learners skim the message, and the headlines tell the story. *As a writer*, you tell what's important and direct the flow of information. *As a learner*, you get the gist of the text and find key information at a glance.

Notice how the Informative headlines give vital information at a glance. Refer to the section "Presenting the Natural Order of Things," later in this chapter to learn when each is appropriate.

Informative: ***Introduction: XYZ Machine Holds Great Promise***
Noninformative: ***Introduction***

Informative: ***Quarterly Inspections Cut Accident Rates by 23%***
Noninformative: ***Report of Quarterly Inspections***

Informative: ***Conclusion: We Need to Conduct Further Tests***
Noninformative: ***Conclusion***

Informative: ***Findings: There Is No Critical Difference Between the Control Group and the Experimental Group***
Noninformative: ***Findings***

SHERYL SAYS

Headlines are important in our daily lives. Case in point: While stopped at a red light the other day, I noticed a sign nailed to a utility pole. It read *MISSING DOG*. I started to wonder what kind of dog? Miniature poodle? Collie? Greyhound? And what color is the dog? If the dog's owner had posted a sign that read, *MISSING DALMATION*, I and everyone else would have known the type of dog to spot (pardon the pun).

Understand the business value of headlines

Once you start thinking about headlines, you'll use them more regularly — in emails, for example. It's the subject line that either grabs the recipient's attention or not. Studies show that 75 percent of email users don't read emails with weak or uninformative subject lines. Have you ever received an email with a subject line, "Meeting"? You probably attend several meetings each day or week and it would be helpful to see more information at a glance. If the subject line read, "Staff meeting at 2:00 in Conference Room A," you would have "gotten" the message even if you didn't open the email. Strong headlines are important in all business communications.

Know the value of SEO in headlines

Headlines help search engines decide whether your writing matches what users are searching for. This may include keywords, proper names, numbers, personal names, product categories, or unique information. While you want to include keywords, don't "keyword overstuff." One way to avoid overstuffing is targeting two related keywords and separating them with colon. In the following example, *iPad* and *Tablet* are two key search words:

How to Buy an iPad: Choosing the Right Tablet

Putting It on the List

If you ever believed in Santa Claus, you know all about making lists. When you prepared your Christmas wish list, you wrote the hottest item as No. 1; the second hottest, No. 2; and so on. If you didn't use numbers, you wouldn't have given Santa any visual clue as to what was most important to you. Santa may have just picked a few things you asked for, and then you'd be disappointed on Christmas morning when your shiny red Jaguar wasn't waiting in the driveway.

When you prepare a shopping list, you list each item but don't use numbers. The list is probably in random order. That's because once you're in the store, you just pick items off the shelf — everything has the same weight (figuratively speaking). Following is an explanation of when to use a bulleted list and a numbered list.

Use bulleted lists

Use bulleted lists when rank and sequence aren't important. Bullets give everything on the list equal value. Always start the list with a descriptive sentence, as you see in Example 5-1.

Following are the fabrication methods for stencils:

- Laser cutting
- Chemical etching
- Electroforming

EXAMPLE 5-1: All bullets are created equal.

Use numbered lists

The bullets that follow show you when to number a list. (Notice I didn't use numbers for this list because one item doesn't have priority over another.)

» **Show items in order of priority.** Doing so gives learners a visual clue that the items on the list are in priority order. (See Example 5-2.)

- **Describe steps in a procedure.** When you describe steps in a procedure, start each numbered item with an *action word* — something for learners to do. (See Example 5-3.)
- **Quantify items.** If you don't number a long list, people may waste time counting the listed items in their heads to make sure the number of items is correct. Also, with a long list, you can say, "Refer to Item 12 on the list." (See Example 5-4.)

Please take care of these issues first thing in the morning. Thanks.

1. Call the ABC Agency to arrange for a consultant for the week of March 15.
2. Ask Jim to prepare his R&D report.
3. Schedule a meeting with everyone involved in the project for the week I return.

EXAMPLE 5-2: Putting first things first.

Following are the requirements for paste formulated for wafer printing:

1. Use a squeegee action to deliver all the stencil aperture contents to the UBM surface.
2. Remove any remaining solder beads with the automated wiping process.
3. Remove oxides from the solder beads during the reflow process.
4. Remove flux residues after the reflow process with mild chemistries.

EXAMPLE 5-3: Take it one step at a time.

Following are the ten key people on the team:

1. Jon Allen
2. Samuel Jones
3. Kim Wong
4. Jackson Pollack
5. Jane Robinson
6. Quincy Adams
7. Dwight Alexander
8. Barbara Geller
9. Pat Lewis
10. Morton Karp

EXAMPLE 5-4: Don't bother counting.

Use parallel structure

Imagine a gymnast in the final tryouts for the Olympics. She gracefully dances along the parallel bars; her eyes are aglow as she looks and smiles at the audience. All of a sudden — oops! — the bars aren't parallel. One bar veers to the left. The poor gymnast falls to the floor. Now imagine your learners, totally absorbed in your document. All of a sudden — oops! — the list or sentence isn't parallel. One component veers off. The poor learners' expectations fall.

Whether you use a bulleted or numbered list, create items that are parallel in structure. That means all elements that function alike must be treated alike. For example, in the parallel bulleted list that follows, all the bulleted items are gerunds — they end with *-ing.* In the nonparallel bulleted list, the first two items end with *-ing*, making the last item (starting with *specify*, which is in italics) stick out like a wart at the end of your nose (regardless of the underlining).

Parallel bulleted list

Effective measures should involve the following:

- Designing and maintaining the facility
- Training the operators and other people in the field
- Specifying security personnel and procedures

Nonparallel bulleted list

Effective measures should involve the following:

- Designing and maintaining the facility
- Training the operators and other people in the field
- *Specify* security personnel and procedures

Punctuate lists

People often get confused as to when to use a colon to introduce a list and when to use a period to end a list. The following demystifies these pesky marks of punctuation. (For more about punctuation, check out Appendix A.)

- **Colons:** Use colons to introduce a list when the words *the following, as follows, here are* or *here is* is stated or implied. However, don't use a colon after a verb.
  - Please consider the following ideas:
  - Please consider these ideas: (*The following* is implied.)
  - The three factors are (In this case, don't use a colon. Just follow the sentence with the bulleted or numbered list.)
- **Periods:** Use periods after each item in a list only when the items on the list are complete sentences.

Avoid laundry lists

When you have too many items on a list, you create a laundry list and learners may just gloss over everything you worked so hard to emphasize. Instead of creating a long list of bulleted or numbered items, break the items into categories. In Example 5-5, you see one long list. In Example 5-6, you see how dividing the list into two logical chunks of information is easier to read and gives more information.

Our global expansion takes us into the following countries:

- Austria
- China
- Hong Kong
- Indonesia
- Malaysia
- Portugal
- Spain
- Sweden
- Thailand

EXAMPLE 5-5: Laundry list of bulleted information.

Our global expansion takes us into the following countries:

Asia

- China
- Hong Kong
- Indonesia
- Malaysia
- Thailand

Europe

- Austria
- Portugal
- Spain
- Sweden

EXAMPLE 5-6: Logical chunks of bulleted information.

WHAT'S YOUR SIGN?

If you want to look savvy in print, use the special signs and symbols that come with your software. If you use (c) for copyright or - - for the em dash, your visual effect will appear amateurish. These "prehistoric" signs date back to the dark ages of typewriters that had limited characters. Following are just a few signs and symbols that are popular in the technical world:

Sign or Symbol	Used for
©	Copyright
®	Registered trademark
™	Trademark
π	Pi
Å	Angstrom
λ	Lambda
µ	Mu
√	Square root

Keeping It Short and Simple (KISS)

We were taught in school how to write. We learned spelling, punctuation, grammar, and more. Perhaps you recall instructors telling you to write a five-page, single-spaced essay on [topic]. If you wrote only four and a half pages, you wouldn't get an A because it wasn't long enough. So you repeated, rehashed, and reiterated what you already said, dragging it out to five pages. That was fine in the academic world, because you needed to satisfy your instructors. But it's not fine in the business world, where *less is more.*

The academic world, therefore, did *not* teach how to write for the business world. Academic writing tends to be long, and it typically has a pre-determined number of pages or words.

Business writing should be brief and to the point. Say what you have to say professionally and politely in as few words as possible. Why? Businesspeople are barraged with information. Reading sprawling sentences and paragraphs is boring, confusing, and overwhelming — making it hard to identify the key points. People may get lost in the weeds, rather than get to the main points. You learn more about keeping it short and simple in Chapter 6.

Presenting the Natural Order of Things

A practice from academia is to put the findings or conclusions at the end. However, that's not necessarily where they belong. When filling out the Technical Writing Brief you determine if your learners' attitude toward your message will be positive, neutral, or negative. That attitude will determine where to put your findings or conclusions.

Scenario: Assume you're asked to conduct a study as to why the Accounts Payable (AP) department is falling behind. You determine that everyone is working overtime, and the only way to increase efficiency is to hire two more people. You prepare a report that will be sent to everyone in the company. However, there are two different reactions:

- People in AP will welcome this as *positive* news. They'll finally get the additional help they need.
- Management will see this as *negative* news. They'll have to include two additional salaries and benefits packages in the budget.

Put the bottom line up front

When you expect a positive or neutral reaction, put the bottom line up front. In the scenario, the AP group will be delighted with the findings. Don't make them rifle through lots of pages to learn the positive information they're hoping for. This isn't a joke where you save the punch line for the end. And don't stop there. Detail in the headline what the findings are, so they can be seen at a glance. Here's the order in which the headlines may appear in the overall report:

Findings: We Need To Hire Two Additional People

Supporting Data

Background

Put the bottom line at the end

When you expect a negative reaction, put the bottom line at the end. In this scenario, you need to build up to *why*. The management team will view this negatively because they're being asked to pay for additional employees. In a scenario such as this, an executive summary is valuable because that's all management may read. (Learn more about writing executive summaries in Chapter 13.)

WRITING EQUATIONS

Even mathematical equations have style. When you write an equation as a sentence, you can give it the form of normal text or break it out on a line of its own. For example, if you want to use an equation within a sentence, you may express it on one line as follows:

The equation you need to know is $z = {}^{1}\!/_{x} = y$

Or you may decide to break it out on a line of its own.

The equation you need to know is

$$z = {}^{1}\!/_{x} = y$$

The following is an appropriate order for these headlines. Notice that the findings are not first. You don't want to call management's attention to the findings until they understand the background and supporting data — the why.

Background

Supporting Data

Findings

TIP

If your report isn't long enough to warrant an executive summary and you find that reactions will be mixed (positive, neutral, and negative), present your report in the order that's appropriate for the highest level people. In the scenario, the bottom line should be at the end with the findings not called out.

A Picture (Pixel) Is Worth a Thousand Words

If a picture is worth a thousand words (or about 600 now, with the rise in inflation), charts and tables are worth gold. They are excellent ways to make your point very effectively. You can gather data and prepare a chart to display your findings, identify opportunities as a result of what visually appears, and update the data to show changes or progress. Many software applications are available to help you prepare graphs in a jiffy. A popular one is Canva (`https://canva.com`), which can turn any novice into a graphic designer.

TIP

If you need to include photos, illustrations, videos, or clip art, there are many sources online. Some are free and some charge a royalty. Check out Unsplash, Freepik, PikWizard, Adobe Stock, Stockvault, Shutterstock, Clip Art, and others.

Keep these tips in mind when you prepare charts and graphs:

- **Include a descriptive title.** Place the title above the chart or graph.
- **Use an appropriate scale.** For example, if your financial range is from $100,000 to $200,000, don't show a scale of $100,000 to $500,000.
- **Create a legend if the chart isn't self-explanatory.** Legends explain the symbols that appear in the chart.
- **Keep the design simple.** Eliminate any information learners don't need to know.
- **Prepare a separate chart or graph for each point.** If you try to squeeze too much information on one graph, you defeat your purpose of making it simple to read.

Include pie charts

A pie chart is like a pizza with wedge-shaped sections. You may order a pizza with 50 percent pepperoni, 25 percent mushrooms, and 25 percent olives. Each section represents a percentage of the total pie, which is 100 percent.

Some people think it's important to begin the most important percentage at the 12 o'clock position and continue clockwise. Others believe that (because people read from left to right) the most important information should be to the left of 12 o'clock and continue counterclockwise. It's your choice. Example 5-7 shows a typical pie chart. Example 5-8 shows a pie chart in three dimensions. (Of course, the 3D pie has more calories.)

Include line charts

A line chart shows trends or the change of one or more variables over time periods, as shown in Example 5-9. Line charts use points plotted in relation to two axes drawn at right angles. Make the axes descriptive and use clear labels.

A slight variation to the line chart is the *run chart*, which shows incidents above and below an established data point. This is evident when you compare Example 5-9 with Example 5-10.

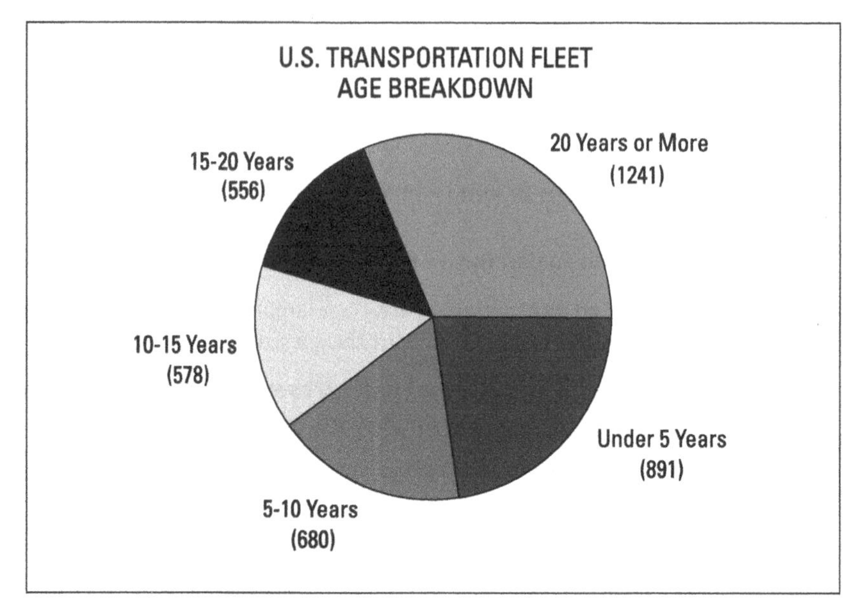

EXAMPLE 5-7: Here's pie in your eye.

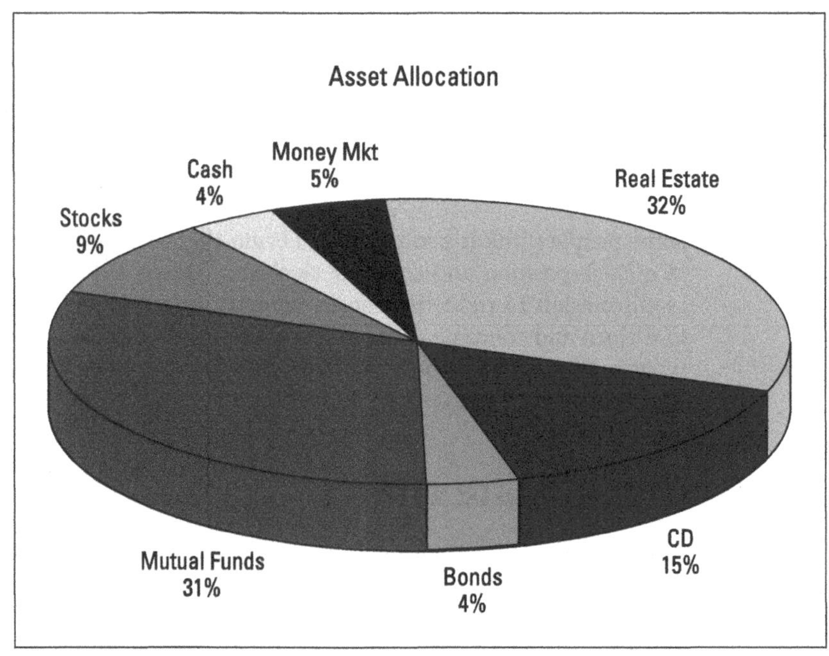

EXAMPLE 5-8: Three-dimensional pie.

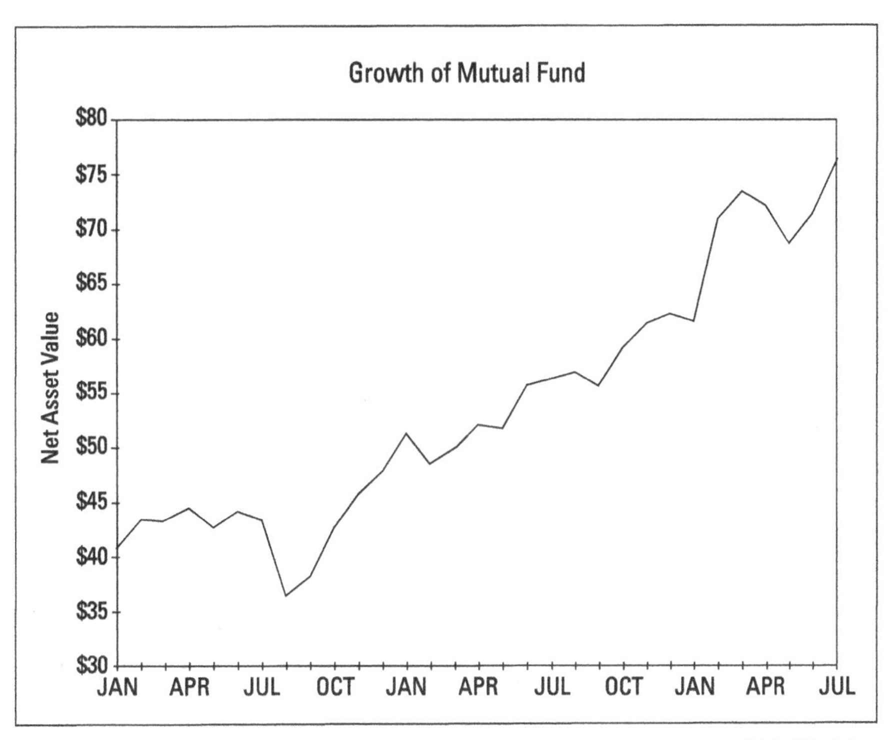

EXAMPLE 5-9: What's my line?

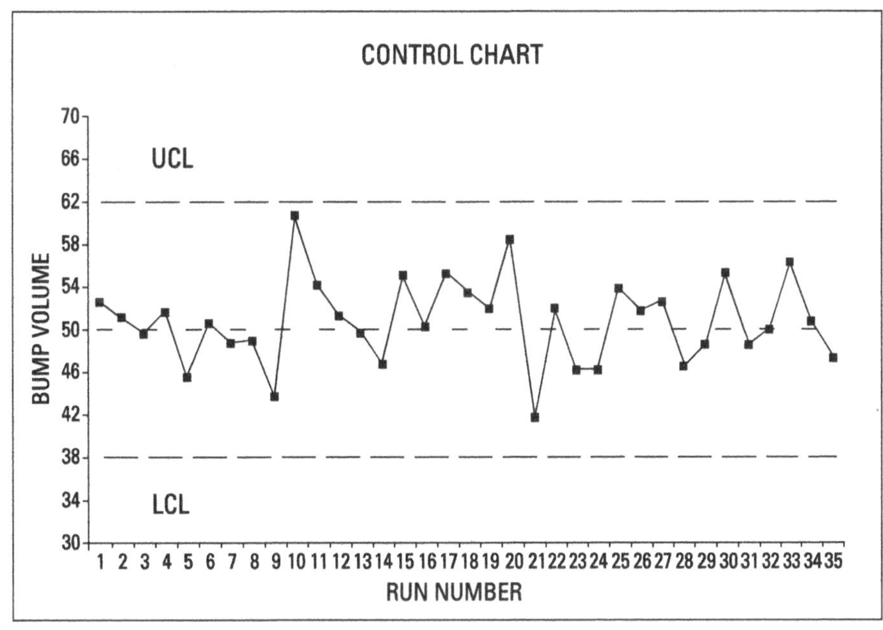

EXAMPLE 5-10: On the run.

Include bar charts

A bar chart (which can use vertical or horizontal bars) shows a comparison between categories, as you see in Example 5-11. Clearly mark the axes. Variations to simple bar charts are histograms, Pareto charts, and Gantt charts, which you use for specific purposes. I describe and give examples of these charts in the following sections.

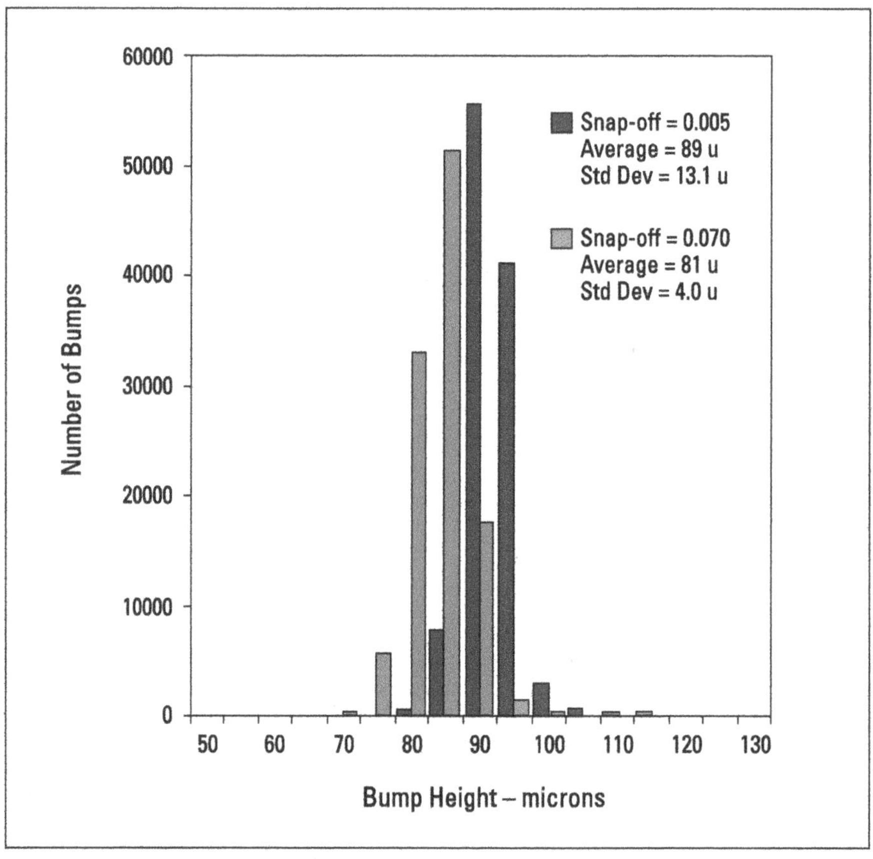

EXAMPLE 5-11: Step up to the bar.

© John Wiley & Sons

Histograms

A histogram shows the relative frequency of occurrence, central tendency, and variability of a data set, as you see in Example 5-12.

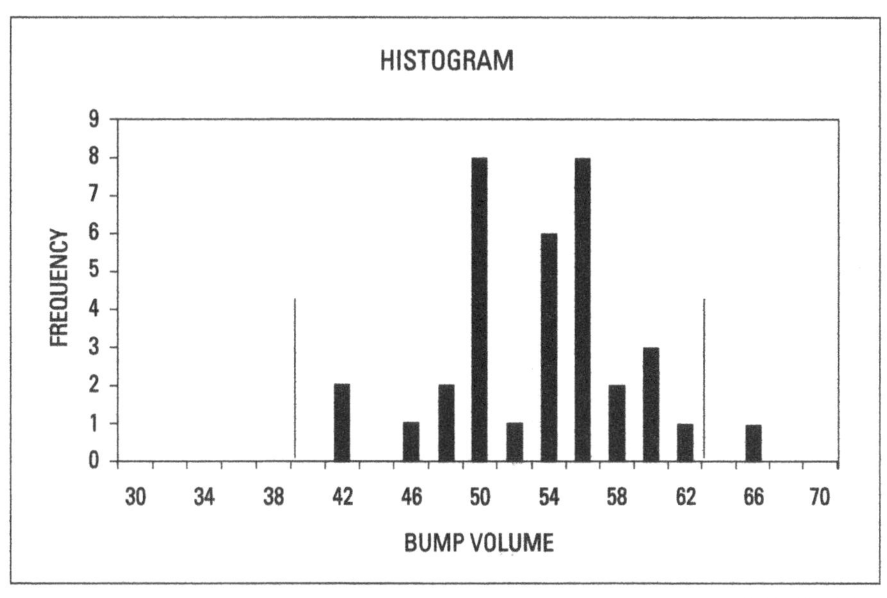

EXAMPLE 5-12: Histogram.

© John Wiley & Sons

Pareto charts

A Pareto chart, shown in Example 5-13, separates vital information from the trivial information. It's based on the Pareto Principle, which says that 20 percent of the problems have 80 percent of the impact.

Gantt charts

The Gantt chart, shown in Example 5-14, is a tool used by management to help coordinate resources and activities. It shows timing relationships between the tasks and subtasks of a project.

Include scatter charts

A scatter chart, shown in Example 5-15, displays a relationship between two variables. It may help pinpoint the cause of a problem or show how one variable may relate to another.

Include flowcharts

Example 5-16 shows symbols used in a flowchart. Example 5-17 displays the major steps in a process using flowchart symbols.

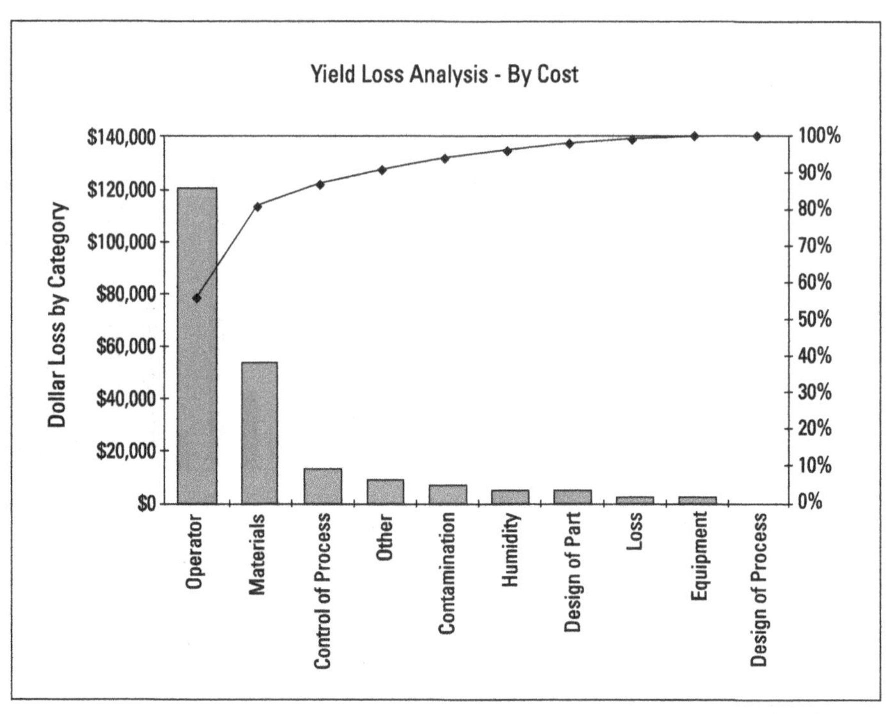

EXAMPLE 5-13: Pareto chart.

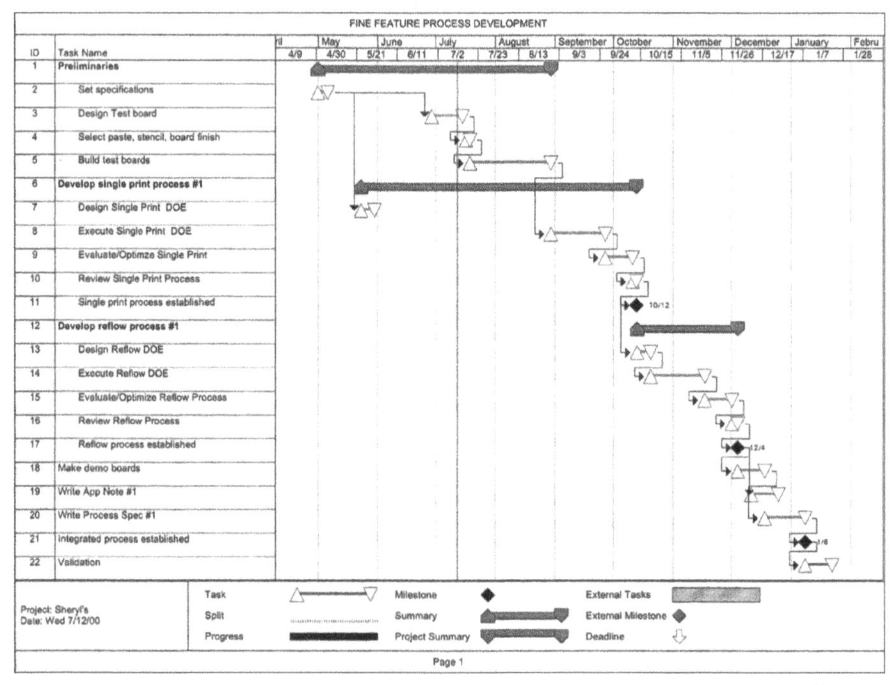

EXAMPLE 5-14: Who's on first?

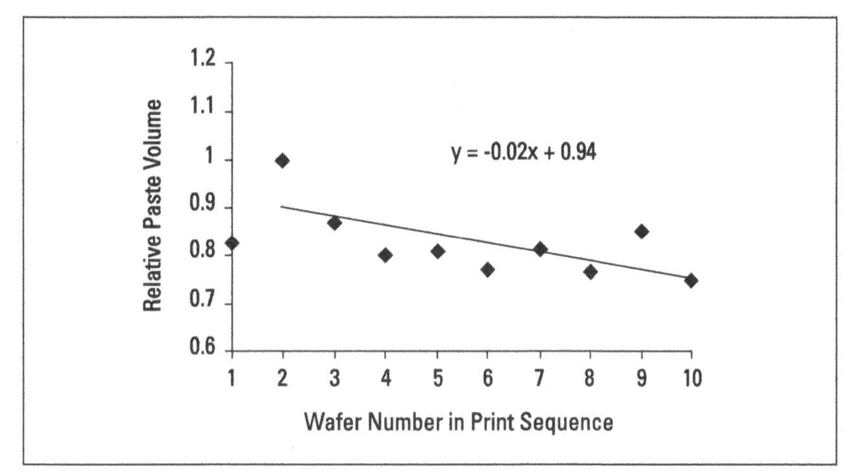

EXAMPLE 5-15: Scattered around.

© John Wiley & Sons

Standard Flowchart Symbols

This symbol...	Represents...
	Start/Stop
	Decision Point
	Activity
	Document
	Connector (to another page or part of the diagram)

EXAMPLE 5-16: Go with the flow.

© John Wiley & Sons

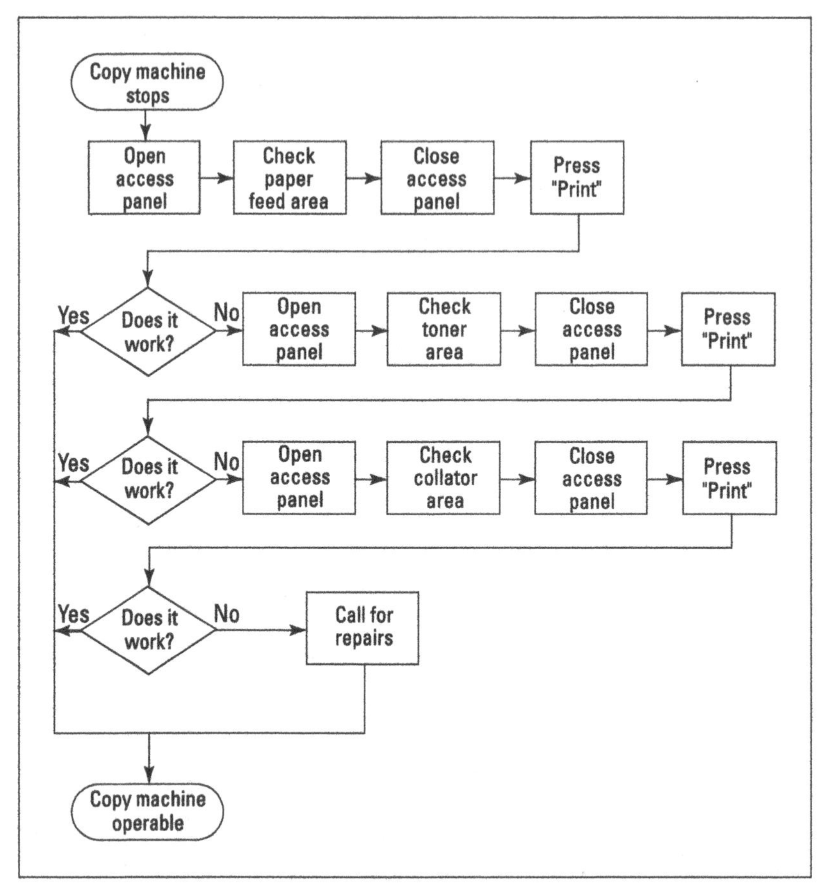

EXAMPLE 5-17: A flowchart in action.

Tabling That Thought

Tables are columns and rows that display specific, related information. Tables carry more information per space than the same amount of text — yet they're often overlooked by technical writers. Find appropriate opportunities to create tables; they have great visual impact. Formal or informal? That's the question. Although there are no hard and fast rules about which tables should be formal or informal, use your judgment based on the formality of your document.

Create a formal table

Separate formal tables from the text with boxed headings, vertical and horizontal rules (lines), and a box, as you see in Example 5-18. If you use more than two or three tables in a document, assign a number to each. Place the table heading above the table. If you need to explain any information, place it below the table as a footnote.

TIP

When you think that the learner may have difficulty following a table across the rows, consider shading every other line, as you see in Example 5-18.

Table 4-5: Complexity Factors

Factor	Low	Moderate	High
Originality required		X	
Processing flexibility	X		
Span of operations	X		
Dynamics of requirements	X		X
Equipment		X	
Personnel	X		
Development costs			X
Processing time		X	
Communication architecture	X		

EXAMPLE 5-18: Setting a formal table.

Create an informal table

Informal tables are extensions of the text and don't have headings or table numbers. Merely write a sentence or two that has ties to the table. For example, in Example 5-19 you see a two-column table about risks that are inherent to a project.

CROSS REFERENCE

Check out Chapter 8 for a great way to create a procedure table when giving step-by-step instructions.

Risks	Pinpointing discrepancies
Staffing	Staffing requirements and the staff available to fulfill those requirements.
Technical	Expected abilities of the technical platform and their actual abilities.
Scoping	Level of functionality and the time and resources available to develop the functionality.
External	Expected behavior of the environment outside the boundaries of the project and those inside the boundaries.

EXAMPLE 5-19: Setting an informal table.

Adding More Value to Your Visuals

The difference between a table and a figure is simple. If a visual element isn't a table, it's a figure. Figures can be sketches (as you see in Example 5-20), drawings, photographs, charts, or graphs — in essence, anything other than columns and rows. If you use more than two or three figures in a document, assign each a number. Include a concise title below or next to the text. Keep figures simple and uncluttered.

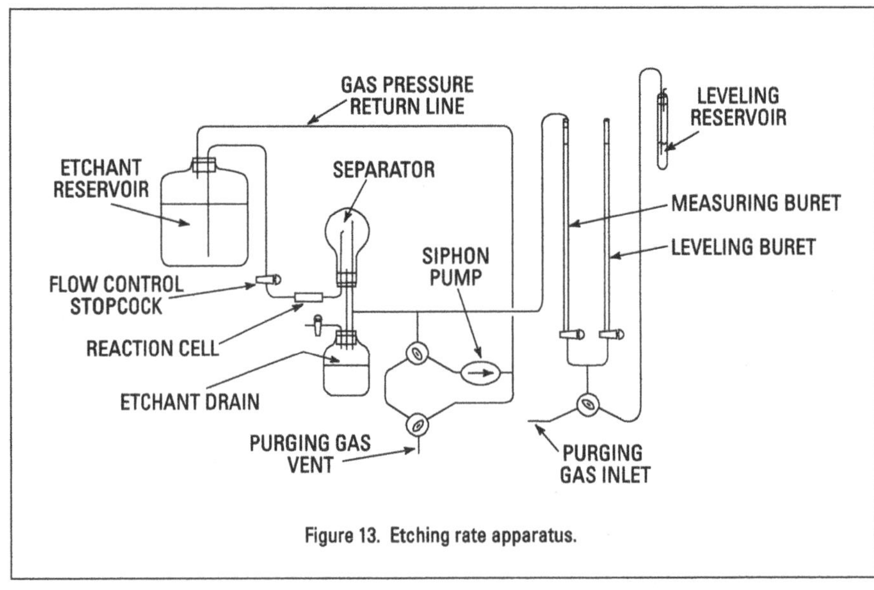

Figure 13. Etching rate apparatus.

EXAMPLE 5-20: The greater scheme of things.

When you use a figure, make sure that it paints an accurate and clear picture. Example 5-21 shows a real-life figure that just boggles the mind. In this case, the thousand words would be better than the picture (to turn a phrase).

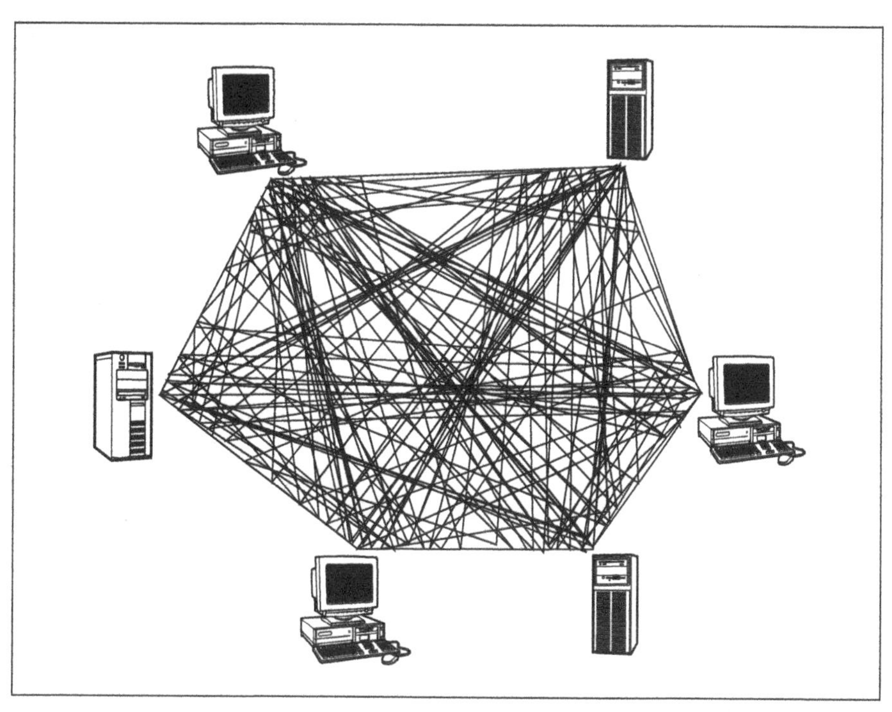

EXAMPLE 5-21: What's the point?

© John Wiley & Sons

Scale for size

When it's important for learners to understand the size of an element, scale the element so learners can clearly envision it.

- **For non-technical learners:** Show something they relate to. In Example 5-22 you see a man's hand holding a small piece of equipment.
- **For technical learners:** Consider a measurement to scale the element. In Example 5-23, you see the element measured in microns. (A *micron* is one millionth of a meter.)

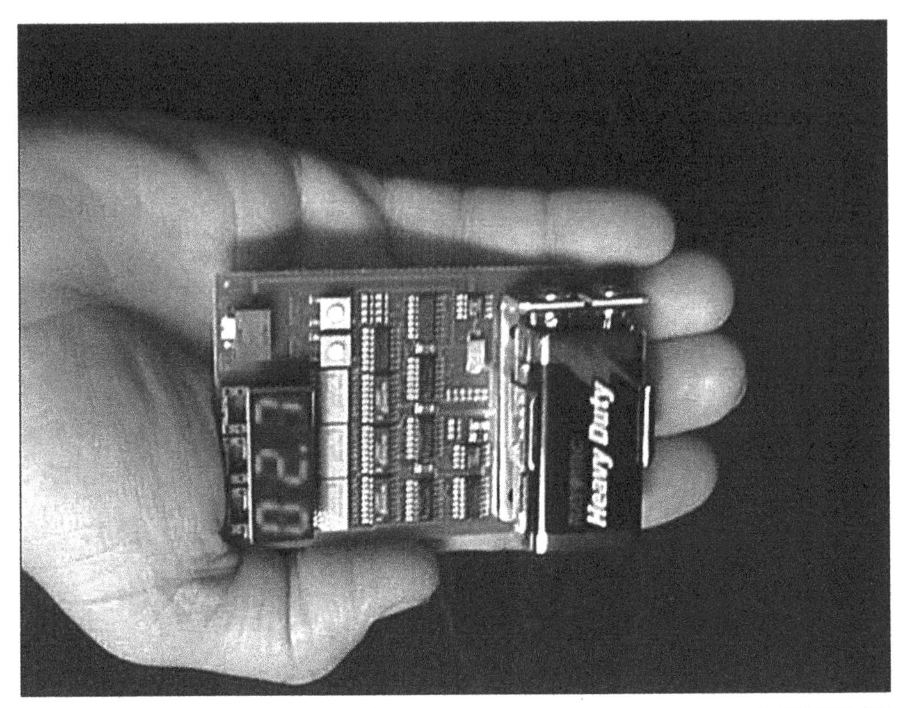

EXAMPLE 5-22: There's no question of size.

© John Wiley & Sons

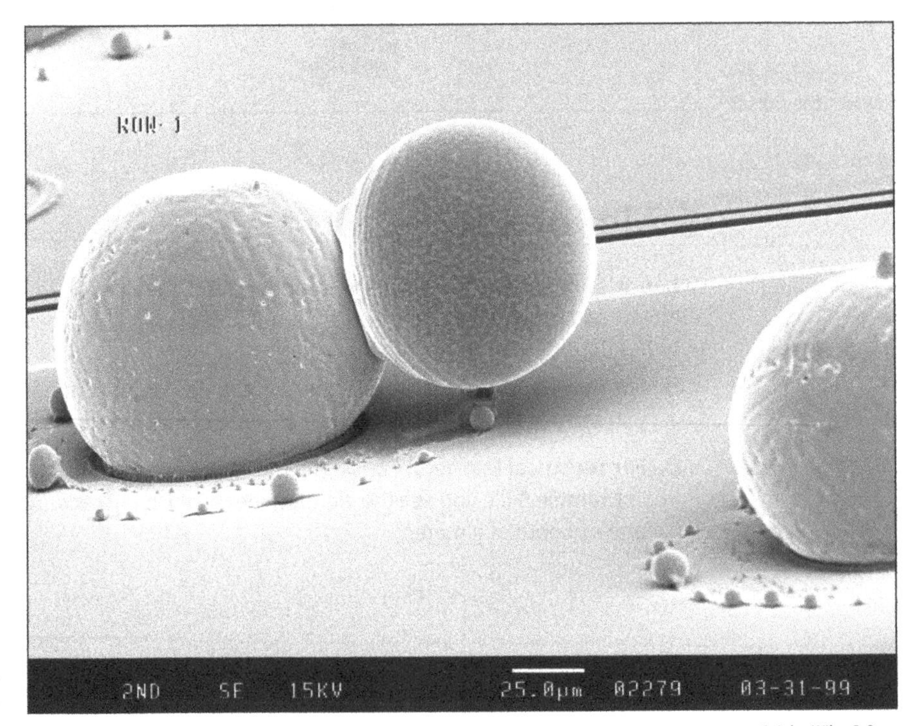

EXAMPLE 5-23: Are these mountains or molehills?

© John Wiley & Sons

Location, location, location

Place graphics — all graphics, not just photos — as close as possible to the related text. If you can't place the graphic on the same page as the text, try for a facing page or the next page. If this isn't possible and the graphic is lengthy, consider putting it in an appendix and cross-referencing it in the text.

What's your visual preference?

The following examples demonstrate three ways to present the same data. They show the effect of a $10,000 investment with compound interest at 10 percent over 30 years.

- **Sentences:** Example 5-24 displays the data in sentence form. This format is cumbersome because all the numbers are jumbled together and it's difficult to read.

The following information represents the growth of a $10,000 investment with compound interest at the rate of 10 percent over a 30-year period: Year 1, $11,046.49; year 2, $12,203.81; year 3, $13,481.66; year 4, $14,891.18; year 5, $16,452.76; year 6, $18,175.51; year 7, $20,078.64; year 8, $22,181.05; year 9, $24,503.60; year 10, $27,069.34; year 15, $44,536.55; year 20, $73,274.92; year 25, $120,557.49; year 30, $198,350.39.

EXAMPLE 5-24: Get out the headache pain reliever.

- **Tables:** Example 5-25 lists the same information in table format. This format is a great way to display the data to someone who crunches numbers and needs the dollar amounts down to the penny (such as an accountant).
- **Bar charts:** Example 5-26 shows a bar chart for the "average Joe" who's interested in envisioning the growth over time.

Growth of a $10,000 Investment with Compounded Interest

Years Invested	Value of Investment
1	$11,046.49
2	$12,203.81
3	$13,481.66
4	$14,891.18
5	$16,452.76
6	$18,175.51
7	$20,078.64
8	$22,181.05
9	$24,503.60
10	$27,069.34

EXAMPLE 5-25: Columns of growth down to the penny.

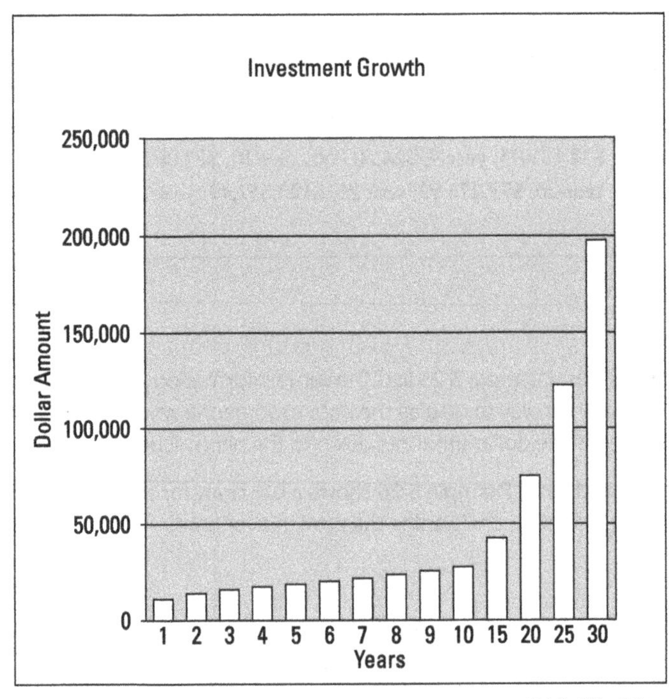

EXAMPLE 5-26: Be wowed by the upward spike.

IN THIS CHAPTER

- » Keeping it short and simple
- » Accentuating the positive
- » Using contractions (or not)
- » Loving the active voice
- » Using diverse and inclusive terms
- » Being clear and consistent
- » Defining terms and acronyms
- » Knowing when to avoid jargon

Chapter 6
Honing the Tone

Get human! Stop trying to speak in a monolithic, generic voice. It's incredibly difficult to write that way, and it's even more excruciating to have to read that kind of content. Why make things hard for yourself and your audience? Just write clearly, in human terms. . . No one believes a monolithic voice, so it undermines your credibility.

—AMY GAHRAN, MEDIA CONSULTANT

Learners aren't tone deaf. They "hear" the subtleties of everything you write. Say it simply. Say it positively. Say it actively. Say it consistently. Your goal is to help your learners understand a concept or process, not dazzle them with overstated language. In essence, take technically mind-numbing concepts and convert them into something easy to digest for the level of your learner.

Giving 'Em a Little KISS

Keeping it simple is the epitome of honing the tone. KISS is an acronym that's used to represent Keep It Short and Simple. (Or as some people like to say, Keep It Simple Stupid.). Either way, the gating word is *simple.* For example, instead of "olfactory impact," write *smell.*

Here's an example from an airline exit-seat card based on federal regulations. It's full of gobbledygook that doesn't add value.

> **Before:** No air carrier may seat a person in a designated seat if it is likely that the person would be unable to perform one or more of the functions under REQUIREMENTS listed below. (33 words)
>
> **After:** To sit in an exit seat you must meet the following conditions. (12 words)

REMEMBER

The essence of good writing is to be concise — shaving everything down to its bare essentials. Concise isn't the opposite of long; it's the opposite of wordy. If information doesn't add value, leave it out. Often, the less you say, the more impact you have.

KISS your technical documents

Clear and simple wording is important in technical documents because technical information by its very nature may be difficult to read. Think about this: It's unlikely you'd say to someone, "As soon as you ascertain the information. . .". You're more likely to say, "As soon as you get the information. . .".

If you surround technical words with simpler ones, your writing is easier to read. The following example illustrates that:

> **Concise:** Please confirm delivery of the HCE exchangers (5%/9%RD) needed by July 1, 2024.
>
> **Wordy:** We wish to request that you notify us if the HCE exchangers (5%/9%RD) will be ready to be shipped and in our hands no later than July 1, 2024.

SHERYL SAYS

Imagine that every word you write costs you $100. That gives you a motivation to cut, cut, cut. Every word that doesn't add to the effectiveness and clarity of your message just wastes your time, the learner's time, and costs money.

Cut to the quick

Why use several words when one will do? Check out Table 6-1 for ways to keep wording simple.

TABLE 6-1 KISSing Your Examples

Use	Instead of
Agree	Came to an agreement
Apply	Make an application
Breakthrough	New breakthrough
Conclude	Arrive at the conclusion
Consensus	General consensus
Consider	Give consideration to
Examine	Make an examination of
Experimented	Conducted an experiment
Investigated	Conducted an investigation
Invited (or asked)	Extended an invitation
Meet	Hold a meeting
Refer	Make reference to
Result	End result
Return	Arrange to return
Save	Realize a savings of
Show	Give an indication
Status	Current status

Accentuating the Positive

Is your glass half full or half empty? When it's half full, you're thought of as an optimist; when it's half empty, a pessimist. (But you already know that.) Let your learners know what they *can and will do*, not what they can't and won't do. Positive words engage the learners' goodwill and enhance your tone. The following sentences are positive and negative ways to send the same message. Notice the difference in tone.

USING CONTRACTIONS — DON'T!

(Oops I just did.) I use contractions in this book because it's a *Dummies* book, not a technical document. Dummies books are written to sound conversational, as if I'm talking to you personally.

The *Chicago Manual of Style* argues that "used thoughtfully, contractions sound natural and relaxed and make reading more enjoyable." Isn't technical writing supposed to be enjoyable as well as informative? You'd think so. However, many people view contractions in technical writing to be unprofessional. So the best advice is to leave contractions in the labor room.

Positive: We expect that you'll be *pleased* with the test results.

Negative: We hope you won't be *disappointed* with the test results.

When you use the negative voice, do it strategically. For example, you may need to call Theodore out by saying, "Theodore neglected to make one-fourth of the corrections," rather than, "Theodore made three-fourths of the corrections."

Fill the glass half full

The following words deliver a positive message. Look for opportunities to pepper documents with them.

Benefit	Bonus	Congratulations	Convenient
Delighted	Excellent	Friend	Generous
Glad	Guarantee	Health	Honest
Immediately	I will	Of course	Pleasant
Pleasure	Pleasing	Proven	Qualified
Right	Safe	Sale	Satisfactory
Save	Thank you	Vacation	Yes

Empty the glass

The following words deliver a negative message. Use them strategically.

Apology	Broken	Cannot	Complaint
Damages	Delay	Difficult	Disappoint
Failure	Guilty	Impossible	Inconvenience
Loss	Mistake	Problem	Regret
Suspicion	Trouble	Unable	You claim
You neglected	Your inability	Your insinuation	Your refusal

Loving the Active Voice

Close your eyes and imagine this scenario: You're vacationing in the tropics with your loved one. The fiery crimson sun is slowly sinking into the distant horizon, and the waves are crashing over the craggy shore. You're sipping a glass of fine wine and affectionately clink your glass with your companion's. That special someone leans over and whispers in your ear, "I love you." Don't those words create a warm, romantic atmosphere?

"I love you" is probably the most wonderful example of the active voice. It's animated and alive! What if that same special someone leans over and whispers in your ear, "You are loved"? (By whom? The dog?) Or worse yet, what if your special someone says, "You are loved by me"? (At that point, you'd probably start checking the dating sites.) The last two attempts at passion are passive. They're dull, weak, and absolutely ineffective.

Bring documents to life with active voice

Voice is the grammatical term that refers to whether the subject of the sentence acts or receives the action. Using the active voice is a major factor in projecting a tone that's alive and interesting. When you write a sentence in the active voice, the subject is the doer (person, place, or thing performing the action). In the following sentence, Dr. Salk is the doer — the discoverer. Can't you almost see him sitting in the lab totally absorbed in his work?

Active voice: Dr. Jonas Salk worked diligently to discover the polio vaccine.

When you write a sentence in the passive voice, the subject is acted upon. Sentences that use the passive voice are often dull and weak. In the following, the vaccine is acted upon. Notice the difference in the impact between the two sentences. In the passive sentence, the actor is gone. This sentence doesn't conjure up much of a vision.

> **Passive voice:** The polio vaccine was discovered by Dr. Jonas Salk, who worked diligently.

Use passive voice strategically

Most of the time, when you use the passive voice, you come across as mealy-mouthed. There are times, however, when you may want to use passive voice because it's more appropriate. Do this for strategic reasons, not as a default. Following are appropriate uses of the passive voice:

- **You want to place the focus on the action, not the actor.**

  The shot was heard 'round the world. (The accent is on the shot, not the people who heard it.)

  The university was established in the early 1950s. (The accent is on the university, not those who established it.)

- **You're hiding something.**

  The tapes were erased. (This is a famous line that came out of the Watergate scandal that led to President Nixon's resignation. The passive voice was used so no one would take the rap for the 18-1/2 minutes that were missing from the tape.)

HOW TO RECOGNIZE A PASSIVE SENTENCE

A passive sentence will always have a form of the verb *to be* in any tense: is, are, was, were, will be, had been, has been, must be, should have been, will have been, is going to be, and more. When you see any of those words, note whether the doer starts the sentence. You can have a form of *to be* in an active sentence, only when the doer starts the sentence.

- **Active:** Julio was late for the meeting. (Julio, the doer, starts the sentence.)
- **Passive:** The defect has been reported by the engineering team. (The engineering team, the doers, don't start the sentence.)

Looking Through Lens of Social Justice

Words matter. Language is ever-growing and always evolving. Look sensitively at diversity, equity and inclusion (DE&I or DEI)) through a lens of social justice:

- *Diversity* relates to the way people differ.
- *Equity* means fair treatment for all.
- *Inclusion* encompasses creating environments in which all people can feel welcomed, respected, supported, and valued to fully participate.

This includes people of color, underrepresented ethnic backgrounds, people with varying abilities, members of the LGBTQI+ community, and women. Just as language can open up new possibilities, it can also reflect harmful norms that force people into boxes.

Show respect for all

Avoid slang, metaphors, and other words or expressions that have a negative historical meaning. For example, "brown bag" lunches has been a term for bringing a bagged lunch to a noontime event. However, brown bags trace back to "brown paper bag tests," which were used to judge the lightness or darkness of African Americans' skin colors. If you're hosting such an event, call it a *lunch and learn.* Instead of *man hours*, use "labor hours" or "work hours." You get the point, but don't stop there:

- Use gender-neutral terms and pronouns.
- Focus on people, not on their circumstances (such as "person with hearing loss," rather than "deaf person").
- Remove all stereotypes about individuals, religions, cultures, and countries.
- Avoid mentions of politics.
- Include diverse and accessible words and images.

Table 6-2 shows gender-neutral terms worth considering.

REMEMBER

When you speak of someone's job title, don't assume their gender. Judges aren't all males, and nurses aren't all females. If you can, identify the person by name.

TABLE 6-2 **Gender-Neutral Terms**

Use This Term	Not This Term
Ancestor	Forefather
Chair, moderator	Chairman, chairperson
Cinematographer	Cameraman
Delivery person, messenger	Delivery boy
Firefighter	Fireman
Fisher	Fisherman
Flight attendant	Steward, stewardess
Humanity, human race	Mankind
Insurance agent	Insurance man
Letter carrier	Postman
Member of the clergy	Clergyman
Meteorologist	Weatherman
Nonprofessional	Layman
Police officer	Policeman, policewoman
Reporter, journalist	Newsman
Sales representative	Salesman, salesperson
Service technician	Repairman, repairwoman
Spokesperson	Spokesman
Synthetic	Man-made
Worker	Workman

Consider gender-neutral pronouns

Many companies have opted to no longer alternate between the use of "he" and "she" to avoid favoring one gender over the other. Instead, it's better to use the singular "they" when referring to individuals.

Here's an example:

> "Not only does your child get to practice what they learn in the classroom, but they also learn to interact with people of all ages, abilities, and strengths. They figure out what it is to join together with others to complete a goal."

Other ways to avoid gendered pronouns include using plural nouns, repeating the noun, and using the pronouns "one" or "who," when appropriate. When a person's gender and pronoun preference is known or established (such as for historical figures), it's appropriate to use the relevant pronoun.

Consider a sentence reword

Gender neutrality can also be a matter of rewording the sentence. The examples that follow show a number of ways to express the same thing:

Acceptable: Each person did the work quietly.

Acceptable: Each person worked quietly.

Acceptable: Everyone worked quietly.

Acceptable: Each person did their work quietly.

Clunky and no longer inclusive: Each person did his or her work quietly.

Being Clear and Consistent

Following are guidelines for maintaining consistency throughout your documents:

- **Be consistent with wording.** For example, if you make reference to a *user manual,* don't later call it *reference manual, guide,* or *document.* Learners won't know whether you're referring to the same publication or to different ones.
- **Don't replace technical terms with synonyms.** Repeating a word is better than compromising the integrity of what you write. In the following "Compromised integrity" examples, the writer changed "computer networks" to "computer systems." Although *system* is a synonym for *network,* the connotation is different. Therefore, the writer compromised the integrity of the sentence.

  Maintained integrity: The members of the networking group were learning all they could about computer networks.

  Compromised integrity: The members of the networking group were learning all they could about computer systems.

» **Avoid ambiguity.** For example, don't use the words *should* or *may* when there aren't options.

Specific: Don't smoke when operating this equipment. (That's definite.)

Not specific: You shouldn't smoke when operating this equipment. (There's a hint of "maybe.")

» **Be precise with words about locations such as top, bottom, left, or right.** Locations are subjective. In the examples that follow, the person may be looking over the computer from behind, so the switch would be on their left.

Specific: As you face the front of the computer, you see the master switch on the right.

Not specific: The master switch is on the right.

» **Use clockwise and counterclockwise to describe turns.** (It's a good idea to use symbols rather than words because they're easier to recognize.)

Specific: Rotate the dial 45° clockwise to create a seal.

Not specific: Rotate the dial 45° to the right to create a seal.

Define terms and acronyms

TECHNICAL WRITING BRIEF

Our understandings are influenced by our backgrounds. Use language, abbreviations, and acronyms your learners will understand and define any that may be misinterpreted or misunderstood. You can determine this when you fill in the Technical Writing Brief in Chapter 3 and consider your audience.

Here's an example: In her book, *Visual Intelligence*, author Amy E. Herman explains how people thought of SOB:

"Many people see it as a cry of despair. Medical professionals told me SOB means 'shortness of breath,' while maintenance crews claim it's 'son of bosses.' For Texas law enforcement agents, it means 'south of the border. . .' [My favorite is] the mother of a teenage texter, who said that SOB is an acronym for 'son of a bitch.'"

Each observation is unique to our personal experience.

SHERYL SAYS

Several years ago, I was trying to describe computing to my 80-year-old mother. She was having a difficult time understanding the difference between hardware and software. When I told her that "hardware is something you can kick, such as the computer," she got it. Of course, this isn't a technical description, but it's one she related to. I understood my audience.

Who's laughing?

Humor is a sensitive area, so use it cautiously. When handled properly, humor can make a technical subject more enjoyable and easier to understand. Here's how humor is handled in *The Lives of a Cell: Notes of a Biology Watcher*. It's a collection of 29 essays written by Lewis Thomas for the *New England Journal of Medicine*. Throughout his essays, Thomas touches on subjects such as biology, anthropology, medicine, music, etymology, mass communication, and computers.

> "Ants are much like human beings as to be an embarrassment. They farm fungi, raise aphids as livestock, launch armies into wars, use chemical sprays to alarm and confuse enemies, and capture slaves. The families of weaver ants engage in child labor, hold their larvae like shuttles to spin the thread that sews the leaves together for their fungus gardens. They exchange information ceaselessly. They do everything but watch television."

WARNING

If there's the slightest chance that your humor may be misconstrued, avoid it. This is especially true when writing for people for whom English is a second language. What's humorous to you may be insulting to them. Always think of DEI.

When to be a jargon junkie

Jargon is specialized shoptalk that's unique to people in an industry. Technical jargon is a hallmark of a good technical document to learners with a vast knowledge of the subject and the terminology. In these cases, watering down the language makes no sense. Doing so may damage the integrity of the document and insult learners. Again, it gets back to knowing your learners.

SHERYL SAYS

Be certain, however, that your learners know the language. I was on a flight waiting to take off when the pilot announced that takeoff would be delayed because of a problem with one of the lavatories. About 20 minutes later, the pilot announced that all the trucks that could fix the problem (a pump out, I assumed) were busy. So the pilot said we'd take off but shouldn't use the "aft lav on the port." Several people looked flustered. If you aren't an aviator, a sailor, or didn't read *Moby Dick*, perhaps you wouldn't know that *aft* means "rear" and *port* means "left."

IN THIS CHAPTER

» Knowing the difference between proofreading and editing

» Crossing your t's and dotting your i's

» Editing for clarity and flow

» Determining the readability of your document

» Using online readability assessments

Chapter 7

Fine-Tuning toward the Ideal

Socrates was a famous Greek teacher who died from an overdose of wedlock.

—FROM A HIGH SCHOOL STUDENT'S EXAM PAPER
(ANONYMOUS FOR A GOOD REASON)

If you don't think that a misused word can make a difference, look at the quote about Socrates. He died from an overdose of hemlock, not wedlock. (At least most married folks hope that was the case.) Now imagine spending days, weeks, or months writing your document only to have a blaring error pop out like a skunk at a lavish garden party. You'd be remembered for your error, not for the rest of your great writing. Remember that diamonds and faux pas are forever!

What would you think if you walked into a restaurant and saw this menu?

Full Coarse Meal

White Whine

Soap of the Day

Frayed Chicken with Hosted Potatoes

-or-

Baked Zits

Tort of the House

Turkey Coffee

SHERYL SAYS

Here's a classic example of an embarrassing moment because the writer didn't take the time to double check the message: A woman in one of my writing workshops is the director of public relations for a major corporation. She told me of quickly composing an email to hundreds of co-workers throughout the United States, Europe, and Asia. In her haste, this director of public relations left the *l* out of public. Can you imagine her humiliation the following morning when she learned that she'd referred to herself as director of *pubic* relations?

Crossing Your T's and Dotting Your I's

REMEMBER

Although your spelling and grammar checkers can pick up a lot of errors, there's nothing like the human eye to see what is and isn't correct. Don't turn on your computer and turn off your brain. Our brains are designed to fill in gaps and make adjustments. For example, you can probably read the following sentence: *We cna reed wrods whit mixd-up lettres evn whn vowles ar missing.* Of course, you noticed errors in nearly every word but still were able to read it. But would you have noticed an error in just one of the words? Perhaps not. Read slowly. Read deliberately. Read with intention. (Yes, the repetition of "read" is purposeful.)

Following are some tips for finding errors that your computer may not detect:

- **Double-check all names, including middle initials, titles, and company distinctions.** Many people get insulted when you misspell their names. Did you type *Lynn* when it should be *Lynne?* Did you write *Corp.* instead of *Co.?*

» **Double-check numbers.** Did you tell the learner the actual height is 7.43 inches instead of 74.3 inches? Errors like this can be critical.

» **Keep an eye out for misused or misspelled homophones (words that sound the same but are spelled differently).** Did you use *principal* when you mean *principle?*

» **Be on the alert for small words you misspelled.** It's easy to type *of* instead of *if* and not notice the error. We tend to read what we expect to be there.

» **Check dates against the calendar.** If you wrote Monday, June 5, be certain that June 5 is a Monday.

» **Check for omissions.** Did you leave off a vital number or other piece of critical information?

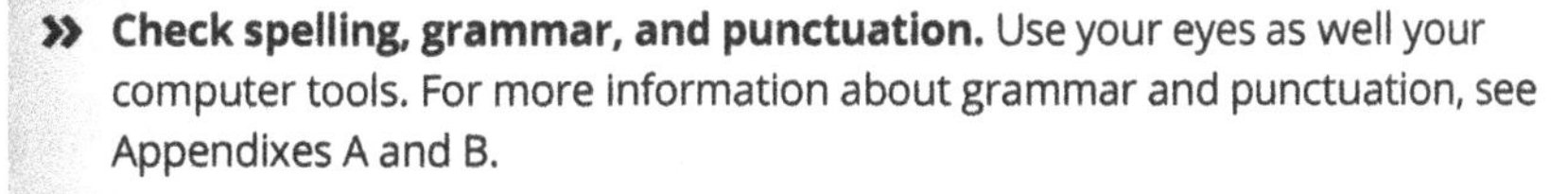

CROSS REFERENCE

» **Check spelling, grammar, and punctuation.** Use your eyes as well your computer tools. For more information about grammar and punctuation, see Appendixes A and B.

» **Print out the message and reread the hard copy.** Why? We're all used to reading the printed word. Therefore, we tend to see errors on hard copy that we didn't notice on the computer screen.

» **Read the text aloud.** (Actually mumble to yourself.) Can you read the document just once and thoroughly understand it? If you can't, reword what you didn't immediately understand.

» **Get a second opinion if the document is critical.** Ask an office buddy to take a look at it.

» **Read from bottom to top and/or from right to left.** Doing so lets you view each word as a separate entity and helps you find errors that you may otherwise miss.

» **Scan the document to see that it looks right.** Is the text aligned properly? Do the examples match the text? Are the graphics in the right place? Are numbered items sequential?

» **Place a ruler under lines of text to help you proofread lengthy material that you copy from paper to your computer.** It's easy to skip a line and never know the difference — especially when the text contains lots of statistical data.

» **Proofread one last time.** After you've completed the final editing pass, read through it one last time. Or even better, ask a co-worker to do this for you.

TIP

When you publish your writings online, double-check the accuracy of all your links. Don't just generate the links; make sure they work.

Editing versus proofreading

There's a difference between editing and proofreading, although both result (hopefully) in a more readable and error-free document. Once you draft the document and the content is correct, it's ready for editing.

Editing refers to amending text by modifying words, sentences, paragraphs, or the general structure of the document. *Proofreading* is the final step before your document is ready for prime time. It refers to the systematic method of finding errors (such as typos) and noting them for subsequent correction. Following are examples of editing and proofreading errors. Both appeared in Ann Landers columns.

> **Editing oversight:** A bean supper will be held on Tuesday evening in the church hall. Music will follow.
>
> **Proofreading oversight:** The rosebud on the altar this morning is to announce the birth of David Alan Smith, the sin of Rev. and Mrs. Julius Smith.

Test your proofreading skills

So you think that you're a proofreading ace. Let's see. Read the following sentence *once* and count the F's in the text. How many are there? You find the answer at the end of the chapter.

> FINISHED FILES ARE THE RESULT
>
> OF YEARS OF SCIENTIFIC
>
> STUDY COMBINED WITH
>
> THE EXPERIENCE OF YEARS.

Editing for Clarity and Flow

While proofreading is about checking surface-level issues, editing gets to the core of the document. It's all about making sure the meaning and ideas are conveyed in the best possible way. Many teams have documents edited before they go to a proofreader to make sure the documents won't embarrass anyone such as the hapless public relations person.

Track changes

Today changes are made by engaging a feature built into Microsoft Word (and other software apps) called Track Changes (sometimes called Revisions or Suggestions). When turned on, this feature keeps track of the edits and distinguishes them from the original text. This is particularly useful for teams with multiple writers and editors. Each person can revise and approve or reject each other's changes. Example 7-1 shows how a document looks when Track Changes are turned on.

prepared a reference card for the data entry folks.

The company had them evaluated by professional reviewers. Both manuals got rave reviews for presentation and ease of use. The company also got great feedback from the engineers and data entry folks. Bottom line: Know your readers and their needs!

<cross-reference>

As you can imagine, knowing who you're writing for is critical to delivering a product that meets and even exceeds expectations. It's so important that I devote an entire chapter to it. For more about identifying your audience, see Chapter 3.

Getting Up and Running

Before you start a project, hold a kick-off meeting for everyone participating in the process: subject matter experts (SMEs), writers, editors, production assistants, reviewers, and anyone else who will give or receive input. Check out Chapter 2 for more details on what you should do when you begin a team project. Following are some of the issues to address at the kick-off meeting:

* **Objectives and scope of project:** Explain the goals of the project. Don't assume that participants know what they are.
* **Development methodology:** This involves the tasks and related activities. It also includes confidentiality issues, sign-off procedures, and audit trails.

EXAMPLE 7-1: Document with track changes.

Use an editing checklist

TIP

Example 7-2 is an editing checklist to review before you finalize any document. It will save you many embarrassing moments! Remember the words of the ubiquitous Ann Onomyous: *The bitterness of poor quality remains longer after the sweetness of meeting the deadline has been forgotten.*

Editing Checklist

- ❒ **My headlines are compelling and will whet the reader's appetite.**
  - ❒ I've included a **key word**(s).
  - ❒ I'm telling a **story**.
- ❒ **The message has visual impact.**
  - ❒ Headlines are **informative**.
  - ❒ There is ample **whitespace**.
  - ❒ **Bulleted and numbered lists**, and **charts and tables**, were used where appropriate.
  - ❒ **Sentences** are limited to 25 words.
  - ❒ **Paragraphs** are limited to 8 lines.
- ❒ **I've reviewed the message for clarity, format, and style.**
  - ❒ The **message will be clear** to my reader.
  - ❒ The message is **logically organized**.
  - ❒ I've **sequenced** the message to keep my reader interested and moving forward.
  - ❒ The **tone** is clear and simple.
- ❒ **My spelling, grammar, and punctuation are correct.**
  - ❒ I used the **spelling checker**.
  - ❒ I checked my **grammar and punctuation**.

EXAMPLE 7-2: Editing checklist.

EDITORS ADD VALUE

When you create a document, you become so familiar with the content, you may take some things for granted. For example, you may write, "Turn the dial to the right." If the dial is on top, the right would depend on which way the user is facing. It would be clearer to say, "Turn the dial clockwise." Editors can be helpful in the following ways:

- Using words learners are likely to understand. For example, saying "smart" instead of "perspicacious."
- Limiting paragraphs to approximately eight lines of text.
- Using clear headlines that tell learners at a glance what the paragraph is about.
- Using visuals so learners can quickly understand complex concepts and processes.

- Reading documents aloud to "hear" how they read. For example, you may have written "is not allowed." If learners slur their eyes over those few words, they may think they read "is allowed." That would give the context a very different meaning.
- Mixing a combination of long and short sentences. Too many long sentences make the content seem burdensome. Too many short sentences give a staccato feeling of stopping and starting too often.
- Using the active voice, when appropriate. For example, in a user manual, don't say "The Enter key should be pressed." Say "Press the Enter key."
- Removing some of the modifiers (adjectives and adverbs). Keep only those that are necessary. Too many cause our brains to do extra work, and our brains already have enough to process.

Determining the Readability of Your Documents

Have you ever purchased a product you needed to assemble? You pull out a user manual printed in lots of languages. There are no visuals, just small print that's difficult to read, even with 20-20 eyesight. So, you get out your magnifying glass and follow step by step. When you finish the assembly, you notice an extra part. You re-read the manual, and there's no mention of that part. What went wrong? No one ever checked the readability, clarity, or accuracy of the manual. It was written, printed, and inserted into the package, probably in haste.

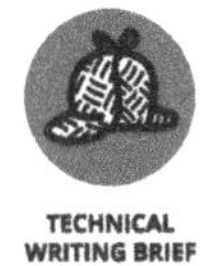
TECHNICAL WRITING BRIEF

Too much technical information is written in this slipshod manner by people with no understanding of the end user. Use the Technical Writing Brief to determine your learner's level of understanding and this won't happen to you.

There are many tools you can use to assess how easy or difficult your document is to read and follow. Microsoft Word and other word processing software can get you off to a good start. Also, search for the words "reliability testing documents" and you'll find loads for free or a fee.

Don't save the best for last

Very few people read technical documents from beginning to end. They read (especially user manuals) to solve problems. Other people don't bother to read the documents but file them away until they need them — especially if they're delivered electronically with loads of others. Ultimately they do read them. And ultimately the documents fail.

REMEMBER

- **High-level people, such as executives, look for conclusion, findings, or executive summaries.** When you're delivering good or neutral news, put that information at the beginning, so learners don't have to rifle through lots of data to find what they really want to know.
- **When you're delivering negative news, build up to it with reasons so your learners will understand.** Otherwise, it may be too abrupt, like punching them in the eye.

Try readability testing

The concept of readability started in 1921, when Edward Thorndike published a book titled *The Teacher's Word Book*. It considered how often difficult words were found in literature and was the first publication to apply a formula to written language. Others followed and built on it — adding average syllables per word, difficulty of words, variety of language, frequency of words, length of sentences, for example. All this calculates a grade level, or readability score.

Then along came the Plain Writing Act of 2010. It requires federal agencies to write clear government communication that the public can understand and use. This was an attempt to make documents understandable when read (or heard) and information in them easy to find. Despite their best efforts, too many technical documents (governmental and others) remain full of incomplete and incomprehensible gobbledygook.

Here are a few statistics you may find interesting:

- Comics are written at 4th or 5th grade levels.
- *The Plain Language Law* is targeted to people with 8th grade reading levels.
- *The New York Times* and *The Wall Street Journal* are written at 10th grade levels.
- Scientific texts have been reducing readability over time, and levels vary widely from one publication to another.

Ramp up your readability

An electronic readability test does *not* replace alpha or beta testing of the product. Share with your "testers" the information you've learned about your learners so they can (try to) put themselves into their mindset:

- » Do the words make sense?
- » Is the document easy on the eyes (or ears)?
- » Are spelling, grammar, and punctuation correct?
- » Did you accurately describe scenarios?
- » Does the information flow logically?
- » Are the steps properly sequenced?
- » Is it complete?
- » Is the document easy to navigate?
- » Do all the links work?

This will give you insights into what information to keep, what to change, what to clarify, where to add visuals, and more. Your goal is for learners to read your documents in as short a time as possible with a clear understanding of your text. Readability does the following:

- » Makes you and your content more credible.
- » Diminishes misunderstanding and allows learners to process information with ease.
- » Improves the likelihood that learners will understand and adapt your thoughts and ideas.
- » Has a positive impact on SEO rankings.

Use online readability assessments

There are several sites that calculate readability. Professionals typically use `https://readable.com`. Also consider using `https://grammarly.com/` for general editing and professional support. It's free. Example 7-3 shows an assessment calculated by several of the leading readability tools.

Flesch Reading Ease score: 53.2 (text scale)
Flesch Reading Ease scored your text: fairly difficult to read.
[f] | [a] | [r]

Gunning Fog: 11.6 (text scale)
Gunning Fog scored your text: hard to read.
[f] | [a] | [r]

Flesch-Kincaid Grade Level: 10.9
Grade level: Eleventh Grade.
[f] | [a] | [r]

The Coleman-Liau Index: 10
Grade level: Tenth Grade
[f] | [a] | [r]

The SMOG Index: 8.9
Grade level: Ninth Grade
[f] | [a] | [r]

Automated Readability Index: 11.2
Grade level: 15-17 yrs. old (Tenth to Eleventh graders)
[f] | [a] | [r]

EXAMPLE 7-3: Readability assessments.

Answer to "Test your proofreading skills": If you counted six, you're right. Most people count three, not paying attention to the word "of," which appears three times.

3 Frequently Written Docs

IN THIS PART . . .

Recognize the value of a well-written documentation (beyond the obvious), master the step-action table, determine FAQs, storyboard for streaming and simulations, and peek into the metaverse.

Understand the types of abstracts and learn when and where they're used.

Learn what a spec sheet is and know what to include on one.

Know the difference between a questionnaire and a survey, learn when to include short and long answers, and benefit from the results.

Get to know your audience, fine-tune your slides, give your audience something to remember you by, and prepare a checklist.

Understand the impact of an executive summary and the anatomy of a report, know what to include, and delve into executive summaries for business plans.

IN THIS CHAPTER

» Understanding the value of a well-written manual

» Using styles and formatting

» Preparing a step-action table

» Knowing what sections to include in the manual

» Determining frequently asked questions (FAQs)

» Scripting an instructional video for streaming

» Writing a simulated learning experience (SLE)

Chapter 8

Writing User Manuals and More

The central paradox of the machines that have made our lives so much brighter, quicker, longer and healthier is that they cannot teach us how to make the best use of them; the information revolution came without an instruction manual.

—PICO IYER, BRITISH AUTHOR

A user manual is also called technical documentation, an operational manual, user guide, instruction manual, quick start guide, training manual, maintenance manual, software manual, or installation manual. Writing a user manual is a big responsibility because it will be used by people who are depending on it for easy-to-follow instructions to successfully complete a task or series of tasks.

Whether the deliverable will be a paper manual, online instructions, streaming video, or simulation — always start by filling out the Technical Writing Brief. You'll find the outline and explanations in Chapter 3.

Understanding the Value of a Well-Written Manual

User manuals are a must for all businesses offering products or services. One that is well written will give customers a feeling of achievement, and they won't be calling customer service for answers to simple questions.

Provide a good customer experience

A user manual is a wonderful marketing tool because it's often a customer's first experience with a product; therefore, so much is linked to its ease of use. A good experience with a manual will cultivate brand faith and loyalty. A perfect example is Apple. From the outset, the company has set a very high bar for providing first-class user guides that are as sleekly designed and easy to follow. Apple understands that everything contained in the product package matters and can influence how users think of them and their products. On the other hand, a manual that's not well written, not easy to follow, and not well packaged leads to frustrated customers. It reflects poorly on the product and the company.

Avoid legal issues

User manuals are particularly crucial when it comes to products that might be harmful or cause major consequences if misused. A user manual should inform users how to use the product safely and warn of potential dangers as a result of its misuse. This helps users handle a product safely and discharges manufacturers from legal liability.

Know what to include

When writing a user manual, here are things to consider including:

- If parts need to be assembled, include a list (with visuals) of everything in the package.
- Prepare a one-page "Quick Start" guide if the instructions are complicated or lengthy.

TYPES OF USER MANUALS

The grim reality of writing user manuals is that no one really wants to read them. People refer to manuals when they have problems or need to figure out not-so-easy-to-understand functions. Manuals can explain how to assemble, how to use, how to fix, and more. They can take a variety of forms: print or electronic media, or a combination of the two. Following are some of the user manuals you may come across:

- Tutorials that are self-study guides
- Do-it-yourself (DYI) assembly kits
- Training manuals used as textbooks
- Operator's manuals written for equipment operators
- Service manuals for technical repairs
- Maintenance manuals for semiskilled technicians
- Repair manuals for service technicians who handle extensive repairs
- Standard Operating Procedures (SOPs)

- Present instructions as step-by-step procedures. (Each step should start with a verb telling the learner what to do.)
- Use lots of visuals throughout. (Images, charts, tables, graphs, screenshots, or whatever will enhance the words.)
- Explain what functions there are and what they're for, not just how to use them.
- Warn of any dangers or anything to avoid. (Make sure warnings are obvious and clear.)
- Write the manual in synch with the product's development timeline — not under pressure of shipping deadlines.
- Make sure you have the experience with product, understand it, and use it as you write.
- Consider the needs of users with disabilities or impairments.
- Test. Test. Test. Have users who aren't familiar with the product use the manual with the product to see if it's easy to understand and follow.

Know your audience

Several years ago, I was asked by a major corporation to write a paper-based user manual. When I filled out the Technical Writing Brief, I realized that 80 percent of the users were data entry folks, and 20 percent were high-level engineers — a ratio of 4:1. There was no way I was going to present a huge tome to data entry folks. They'd either faint or quit on the spot. To meet the needs of this diverse group of learners, I prepared two separate manuals:

- **Engineering user manual:** I wrote the overall user manual for the engineers. It contained all the bits and bytes and nuts and bolts. This manual came to 500 pages, which I broke into major chapter headings, subheadings, sub-subheadings, and so forth. The company housed these svelte manuals in three-ring binders and needed to provide weight training belts for heavy lifting.
- **Data entry user manual:** I wrote a separate 30-page manual that was distributed to the data entry people. It included all the tasks they needed to perform and a step-by-step approach to performing them. I had this manual bound separately (with its own cover) and three-hole punched so it could be inserted into every engineering manual as well.

The company had the manuals evaluated by professional reviewers. Both manuals got rave reviews for presentation and ease of use. The company also got great feedback from the engineers and data entry folks. Bottom line: Know your learners and their needs!

As you can imagine, knowing who you're writing for is critical to delivering a product that meets (and even exceeds) expectations. It's so important that I devote an entire chapter to it. For more about identifying your learners, see Chapter 3.

Getting Up and Running

Before you start a project, hold a kick-off meeting for everyone participating in the process: subject matter experts (SMEs), writers, editors, production assistants, reviewers, and anyone else who will give or receive input. Check out Chapter 2 for more details on what you should do when you begin a team project. Following are some of the issues to address at the kick-off meeting:

- **Objectives and scope of project:** Explain the goals of the project. Don't assume that participants know what they are.

PAPER STILL HAS POWER

Although online instructions have become the norm, instructions are used in many different environments: indoors or outdoors, with good or poor lighting, in safe or dangerous settings, in areas with poor or no WiFi, and so on. Therefore, there are still needs for paper manuals. I recently had a centralized dehumidifier system installed in my home. It came with a thermostat that was very complicated to install and use. (It did everything except remind me of the birthdays of my loved ones.) The installer was having a problem trying to sync the unit (in the basement) with the remote thermostat in the bedroom. There was no paper manual. He called the company's customer service and was on hold for nearly 20 minutes before he was disconnected. After a few choice words he mumbled, "Whatever happened to the good old days of paper instructions? Boy, do I miss them!"

Here are some basic guidelines to ensure your paper manual will survive actual use:

- Ensure the manual can lie flat on a work surface when opened. This will help to determine the binding.
- Consider whether users need to hold the manual and work at the same time. This will help to determine the size.
- Provide durable covers and pages.
- Consider whether the manual needs to resist water, oil, dirt, grease, and so on.

» **Development methodology:** This involves the tasks and related activities. It also includes confidentiality issues, sign-off procedures, and audit trails.

» **Ground rules:** The ground rules cover everything from exchanging information to handling problems.

» **Roles and timeframes:** Discuss people's responsibilities and how much time they have to complete their tasks.

» **Milestones:** Manuals often involve rounds of drafts and field testing. Include all the critical milestones. Plug in dates and identify the people responsible.

There are many strategies for writing. The trick is to find what works for you. I'll be sharing approaches that work for me — what I teach during the technical writing workshops I facilitate — and what works for the hundreds of people who've learned and incorporated these strategies into their writing. Whatever strategy you use or develop, some things remain the same. Always keep the user in mind.

Instructional materials must be clear and concise. Use graphics. Most important of all, experiment, have fun, learn, and engage in an ongoing dialogue with your computer; it's your friend.

Determining Style and Format

User manuals must be brief, yet detailed enough to give accurate, easy-to-follow directions and reasons for doing so. In other words: maximum information in as few words as possible. Look at the following before and after documents and notice the difference in language and usability.

Example 8-1 is the *Before* document. It has lots of unnecessary and confusing numbering. It doesn't say why it's important to follow the guidelines. And it's loaded with gobbledygook that users don't need to know and don't care about.

JOHNSON SPACE CENTER (MSC) HANDBOOK

101.1 Policy

It is the basic policy of JSC to take all practical steps to avoid loss of life, personal injury or illness, property damage or loss, or environmental loss or damage.

101.2 Goals and Objectives

101.2.1 Goals

a. To achieve a successful and unified occupational safety and health program while accomplishing JSC's objective for excellence in human space flight.

b. Of equal importance, to become a nationally recognized center for excellence for occupational safety and health. This excellence will also be a prominent feature of JSC's environmental protection and emergency preparedness program.

101.2.2 Objectives

JSC shall comply with all applicable regulations and standards, including those of the Occupational Safety and Health Administration (OSHA) and the Environmental Protection Agency (EPA). By exercising flexibility and creativity in striving for excellence, JCS will go beyond the minimum requirements of the regulations and standards to provide the best feasible protection for workers at JSC and the environment within the constraints of available resources.

....

Courtesy of Johnson Space Center, NASA

EXAMPLE 8-1: This Before document is not effective or clear.

Example 8-2, the *After* document, is clear and concise. It gets right to the point, gives information at a glance, and provides learners with a good reason to follow the instructions.

JOHNSON SPACE CENTER (MSC) HANDBOOK

WHY THIS IS IMPORTANT

This could be you. . . a hypergol technician didn't follow requirements and caused a major fuel spill. He was burned. He was using a tool that fell to the floor causing two employees spilled a caustic battery electrolyte on their hands. The batteries hadn't been through qualification testing. There were no requirements to prevent the technicians from working with unqualified batteries. That's why this it's important to follow the guidelines in this handbook.

WHO MUST FOLLOW THIS HANDBOOK

This handbook applies to anyone at JCS headquarters or JCS field sites, unless exempted.

If you...	***Then you must follow...***
Are a federal employee	This handbook unless you work at a site that involves unique military equipment and operations.
Are a JSC contractor	This handbook as called out in your statement of work.
Work at a JSC remote site (such as the White Sands Test Facility) as a civil service employee or contract employee	All chapters that don't exempt you and local requirements that meet the intent of any chapter that exempts you.
Are a non-NASA or non-contract employee	This handbook while on JSC property.

EXAMPLE 8-2: Clear and concise with key information appearing at a glance.

Courtesy of Johnson Space Center, NASA

ADDING IMAGES TO YOUR INSTRUCTIONS

When writing instructions, blending text and images/screenshots gives learners additional clarity and makes your instructions easier to follow. That's analogous to looking for a specific location while driving. For example, when you're told your destination is next to The Home Depot on the left side of the road, spotting The Home Depot sign is a valuable visual indicator that you've arrived. When you give learners good visual clues in written instructions, they'll know they are on the right track. If the image isn't something your company generated or isn't in the public domain, you may need to obtain permission for its use.

Taking Each Step, Then Acting on It

A step-action table, as you see in Table 8-1, breaks tasks into easy-to-follow, step-by-step instructions. This works well whether the instructions are paper or online. Here are some guidelines:

- » Create a title for each process or procedure.
- » Start each step with a verb, something learners should do.
- » Use clear, understandable language.
- » Include within the Action column any notes, results, screenshots, graphics, photos, if-then tables, or anything else that may further clarify the action.
- » Keep the language consistent. For example, don't say user manual, and then call it a guide or handbook.

Any step-action table can be extended with additional columns. For example, if you're writing instructions for a very large project, perhaps you may include a column for "date completed," "responsible party," or whatever is appropriate.

TABLE 8-1 **Start-Up Procedure with Easy-to-Follow Instructions**

Step	Action
1	Email your company's sub-contractor supply list in Excel format to ABC Company.
2	Go to the ABC Company website (`www.abccompany.com`) and fill out the Contractor sign-up form; enter your company information. Result: ABC Company will link your contractor account to the privatized subcontractor/supply list you sent us.
3	Check the box to show you agree that a standardized letter will be sent to your entire database. Notes: • This will notify all your contacts that your company is taking advantage of the ABC Company services to help contractors become more effective during the invitation-to-bid process. • This will publicize your company while you gain the interest of subcontractors.

SHERYL SAYS

My husband, who is technically savvy, wanted to update the GPS in his three-year-old car. (I'll leave the name of the car manufacturer out to protect the guilty.) Despite the poor instructions, he was able to remove the GPS Navigation SD card. Then the frustration erupted.

When the paper manual that came with the car didn't include adequate instructions, he turned to the Internet. When the Internet didn't provide adequate instructions, he called customer service. When the customer service rep didn't give him adequate instructions, he drove 45 minutes to the dealership. When the technician at the dealership couldn't help, he at least learned what to do. He needed to create a profile before he could perform the GPS update. Neither the paper instructions, the Internet, nor the customer rep told him that. He spent too many frustrating hours because of this major omission in the instructions.

Lesson learned: Whenever you prepare instructions, if there's something that must happen prior to beginning the process, be sure to include a section titled "Before You Begin."

The devil is in the details

You must write instructions with clarity and keen attention to detail. Never assume your learner will read between the lines or read your mind. For example, the following figure shows nine dots. Join them together with four straight lines without lifting your pen(cil) from the paper. Try it.

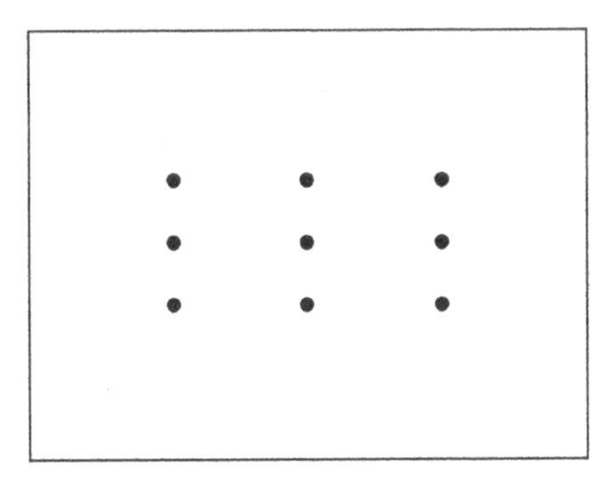

Did you do it on the first try? If not, that's because the instructions are vague. When you write vague instructions, your learner can't complete the task successfully. Look at the end of this chapter for how to connect the nine dots with four lines.

Check out the contents of the box

Here's a scenario you may relate to: You buy a new gas grill at a greatly reduced pre-season price. You can't wait to get home to smell the aroma of your juicy red steaks sizzling on the grill. When you open the box, however, you notice that the grill is completely disassembled. The instructions say, "This is a do-it-yourself

(DYI) project that any ten-year-old can put together." It looks as if the sizzling will have to wait until you assemble the grill. You spread out the grill parts on the ground but can't tell whether you have all the pieces. The instructions don't tell you what parts are enclosed. You won't know whether you have the right parts until the grill doesn't work or you have parts left over.

Lesson learned: Always tell your learner what's in the box (or packaging) so there are no surprises. It's a good idea to list all the parts and show photos or drawings of any parts your learner may not recognize.

Writing for Between the Covers

When you prepare a lengthy user manual, there are sections to supplement the body of text. The following sections outline what to include.

Prepare a table of contents

Whenever the manual is more than 15 pages, include a table of contents to help learners easily find what they need. Use leaders (. . . . . . .) to connect the subjects to the page numbers so learners can run their eyes across the page.

TIP

Some writers use internal chapter numbers. For example, Chapter 1 is numbered 1-1, 1-2, 1-3, 1-4; Chapter 2 is numbered 2-1, 2-2, 2-3, 2-4, and so forth. This is different from consecutive numbering (1, 2, 3, 4). Use internal chapter numbers when you anticipate updating one or more chapters. Doing so keeps you from renumbering (and reprinting) the entire manual.

Append appendixes

Appendixes allow you to include helpful information that would otherwise clutter, break up, or be distracting to the main body of the text. That may include tables, graphs, charts, images, resumes, or more.

Generate a glossary

If there's a hint that your learners may not understand many of the terms or acronyms, include a glossary at the end. If you define a word or acronym the first time it appears in the book, that may not be adequate because learners won't read your manual from cover to cover. They'll use it for reference when they have a question and may have missed the first mention.

Itemize an index

Not all manuals need an index. The content of the manual is the gating factor, not the length. Put yourself in the mindset of the learner and ask: If I were reading this document, would an index be helpful? If you think an index would help you, include one. *Be very sensitive to the logical search words that the learner may look for.*

I heard a story of a woman who bought a new car. She drove about 25 miles when she ran over some glass and got a flat tire. The woman wanted to change the tire herself and pulled out the owner's manual. She checked under the following letters in the index:

f for flat

t for tire

j for jack

c for change

The woman couldn't find the information and couldn't believe that the manual neglected such an important entry. She finally gave up and phoned the American Automobile Association (AAA). When she returned home, she began checking every index entry. Yes, there was an entry for changing a flat tire. It was under *h* for "How to change . . ."

MAKING A LIST AND CHECKING IT TWICE

After you write the document and *before* you send it for testing, check it for basic proofreading and editing errors (see Chapter 7). Then answer the following questions:

- Is the manual complete?
- Do I need a getting-started section?
- Is it well organized?
- Is it technically accurate?
- Did I explain terms or abbreviations that learners may not understand?
- Did I use consistent language?
- Are the examples in sync with the text?
- Are the examples placed where they belong?
- Do the subjects and page numbers in the text match the table of contents and index?

Testing, Testing, 1-2-3

You won't know whether your manual is accurate and understandable until you have it tested extensively. Ask your testers to keep a detailed log of everything that doesn't work, is unclear, or is wrong. Following are a number of approaches:

- Get a SME to review the manual for technical accuracy.
- Test the manual in typical user environments, if possible.
- Ask a novice user to read the manual to make sure that it's accurate and easily understood. A novice often finds a need for explanations or instructions that you or more sophisticated users take for granted.

Determining Frequently Asked Questions (FAQs)

Most consumers try to resolve their issues before reaching out to customer support. That's why online FAQs are good business. They decrease the burden on customer support reps so the reps can focus on more pressing issues. Also, well-done FAQs are a unique opportunity to address the concerns of potential users, removing obstacles on the path to purchase.

How do you know what questions to include? Scour your inbox and ticketing system, and quiz customer support reps to learn what callers typically struggle with. Social media channels can be a useful source of customer frustrations as well. Also, learn what doubts people had before making a purchase.

- Write questions from the user's point of view, as if the user is talking to a person. Example: "Can I *use*. . . ?"
- Start with a "yes" or a "no" answer when you can. Then follow it briefly with the answer. Example: Q. Do you test on animals? A. No. We do not test on animals.
- Categorize your topics. They may include Shipping, Updating Your Order, Sizing and Fit, Order Issued, Return Policy, Product FAQS, and more.
- Make the FAQs searchable so users aren't forced to scroll through irrelevant questions to find what they're looking for.

- Update your FAQ section whenever you roll out new features, products, or updates — or whenever you learn of the same questions that keep popping up.
- Refer to relevant whitepapers or websites.

Preparing an Instructional Video for Streaming

If you ever wanted to learn a craft, cook like a pro, or just about anything else, chances are you've watched a YouTube video. Today people watch how-to videos when they need to get instructions or solve a problem. Instructional videos must be easy to follow, visually dynamic, and speak to the target audience's needs. When writing the script, prepare the instructions in chunks of information because there's no guarantee people will watch beyond their immediate need.

As a rule of thumb, prepare a script for a video that will run for no more than five minutes, unless there's a reason for a lengthier one (such as for complicated technology). A series of short videos, chunked into chapter-like segments, is more useful than a very lengthy one. Here are a few tips:

- Determine if your video will be animated or live action.
- Decide on the length and number of segments.
- Know your budget.
- Think in terms of images and tell a story. (Yes, even instructions can be in story form.)
- Prepare a script (such as the one in Table 8-2).
- Use humor cautiously and only when appropriate.
- Use an experienced narrator if the budget allows. If the narration isn't clear, it will reflect poorly on your product.
- If you're including music, make it appropriate for the industry. (Use license-free music or pay for the rights. Otherwise it will be removed from social media or major streaming services.)

One difference with scripting for videos is to be aware of how your words will "sound," rather than how they'll "read." For example, do you stumble over a word when you say it but gloss over it when you read it? And never underestimate the time it takes to complete a video production. On average, expect a three-minute video to take several days to write and polish. Why? Just like in the movies, there will be edits, retakes, more edits, and more retakes.

Start with a script

In order to write a script, you'll need to fill out the Technical Writing Brief (such as the one in Chapter 3) so you can understand with your audience and empathize with them. The Brief will give you a deeper understanding of how to communicate with your audience by knowing what motivates them. You'll be able to create a story to engage them with succinct, relevant, and compelling content which results in what you want them to do, think, feel, or learn. The key is translating these core messages into a simple story.

Everything you tell and show will depend onthe objective your video needs to achieve. It's not just about what you want to say to your audience: It's about saying it in a way that resonates with them and makes it memorable. *Stories resonate and are memorable.* Think about it; your video is a story. It has a beginning, a middle, and an end. That's a story. Incorporate into your overall story, real-life stories that relate to the topic.

Here's how to complete a script, such as the one in Table 8-2 that shows the opening of a first-round draft:

Column 1: In the Tell column, include the words to be narrated. The draft can be short notes of key things to say. Complete this column before working on the Visual column, unless you're clear about the visuals. You'll undoubtedly have several rounds before the final version, which will be verbatim with all the visuals in place.

Column 2: Your video isn't a book. The story in video is told not only through words, but through graphics, music, sound effects, voiceovers, and anything else that will add "voice" to the narration. Some of this can be filled in when writing the draft, if the visuals are clear in your mind at that time. If they're not, either leave the frame blank or use TBD (to be determined).

Column 3: The Duration column addresses the length of time for each segment. Start filling this in while writing the first draft to get an idea of the video's length.

TABLE 8-2 Sample Video Script: How to Fly an XXX Aircraft

Tell	Visuals	Duration
	Aerial view flying over New York City Accompanying music: TBD	10 seconds
Welcome to the XXX Aircraft. This exciting journey will take you. . . (Objectives and outcomes)	TBD	2 minutes
Before we begin with your journey, let's examine some of what. . . . (Describe in detail what learners will see.)	Show the cockpit	10 minutes
Now sit back and enjoy the ride.	TBD	12 minutes

Choose a video host

While YouTube is probably the most popular online video host for general instructions, you may need a private, more secure, or customized alternative. There are plenty of online platforms to choose from such as Vimeo, Brightcove, Sprout Video, and others. Optimize your video for all devices. Be sure to add closed captioning. Make it easy for users to view your instructional video in any learning environment.

Creating a Simulated Learning Experience (SLE)

Otherwise known as Simulation Scenario Writings, SLEs are techniques that immerse learners in tasks or settings as if they were in the real world. This is a form of Artificial Intelligence (AI) that has barely scratched the surface. When the first edition of this book was released 20 years ago, SLE was science fiction. Can you even imagine where AI will take us in the next 20 years?

SLEs are now being implemented in many industries, including automotive, manufacturing, construction, aircraft, healthcare, extractive industry, military, law enforcement, education and real estate. They're used for learning technical and functional expertise, problem-solving and decision-making skills, interpersonal and communication skills, and team-based competencies. SLE makes instructions smarter and real. Example 8-3 shows a soldier using virtual reality (VR) glasses for combat simulation training.

EXAMPLE 8-3: Combat simulation training.

TimeStopper/Adobe Stock

SLE's have ushered in Augmented Reality (AR), which superimposes digital information on devices such as smartphones and smart tablets. Virtual Reality (VR) places users in a virtual environment through head-mounted devices. Mixed Reality (MR) combines the two, and Extended Reality (XR) is an umbrella term that covers the various experiences that enhance our senses.

Apply SLE to simulations

SLE is being used for designing, training, teaching, and more. One application is medical students learning about the human body by applying AR to cadavers. Pilots learn to handle in-flight emergency situations using VR simulations. And car manufacturers use AR and VR technologies to design cars and create HUDs (heads-up displays) for safety and performance. Here's a glimpse of two projects that are underway:

» The Defense Advanced Research Projects Agency (DARPA) has issued a multi-million contract to a team building an AI system to scan instruction manuals and convert them into instructions for AR systems.

» Lockheed Martin uses AR goggles to assemble its space systems for NASA so that technicians can see relevant information and instructions in the space around them.

REMEMBER

Regardless of the sophisticated tools, remember that "content is still king." Authoring tools for these new technologies are in the early stages of development, but are rapidly evolving. Content that was already prepared for more conventional training is being migrated into simulated experiences.

Remember that *instructions are instructions* regardless of the presentation format. The basic guidelines of clear and simple writing always prevail. This new frontier for technical writers calls upon the basic skills plus a sophisticated mindset. Writers need to think of the written content and align graphics and simulations with real-world objects to be implemented in the world of AI.

Imagine what's next: The metaverse

Many elements we know today as AI and VR are in the embryonic stages of the metaverse (meta + universe). Companies such as Microsoft, Google, Amazon, and others are pioneers in this new "space" frontier. The potential for technical writing is vast.

But what is the metaverse? It's an integrated network of 3D technology that takes us into virtual worlds — beyond simulations. People can enter these worlds as avatars through VR headsets and move about using eye movements, feedback controllers, and voice commands. Components of metaverse technology have already been developed within online video games. Proposed applications for metaverse technology include improving work productivity, interactive learning environments, e-commerce, mass-audience interaction, real estate and fashion, and so much more.

We're a long way from seeing the full potential of the metaverse or understanding precisely where it will take us. Just remember, in the early part of the 1900s, the Internet was science fiction.

Join-the-Dots Brain Teaser

The instructions in "The devil is in the details" earlier in this chapter were vague. If they were detailed, they would have told you where to start and another would have told you that you don't need to stay within the perimeter of the square. Here's the solution:

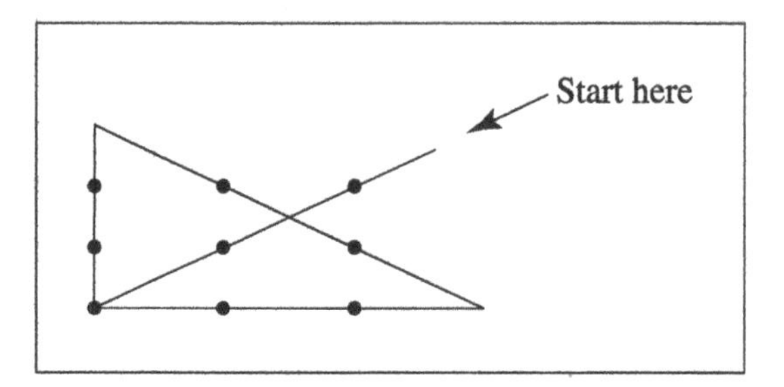

IN THIS CHAPTER

» Understanding the types of abstracts

» Knowing how and when abstracts are used

Chapter 9
Preparing Abstracts

Much wisdom often goes with fewest words.

—SOPHOCLES, GREEK TRAGEDIAN

You write an abstract to condense the key issues of a longer document. This is similar to the trailer of a movie that helps viewers determine whether they want to see the full feature. You don't write an abstract for all your documents, only for documents that are very long or highly technical. Abstracts can become critical segments of journal articles, theses, book proposals, grant applications, and more. Keywords are important because online databases use abstracts to make it easier for people to research a particular subject or find a body of work.

TIP

An *abstract* is a subset of an entire document, so you can't write it before you complete the document.

Types of Abstracts

There are a five types of abstracts: descriptive, evaluative, critical, highlight, and informative. I discuss descriptive, evaluative, critical, and highlight abstracts briefly and informative abstracts in greater detail because they're most commonly used and have a more widespread purpose.

- **Descriptive abstracts:** In about 100 words, they provide an overview of the content, dealing with major points and research methods — not findings, conclusions, or recommendations. Learners need to view the longer document to garner that information. This is helpful when you need to produce an abstract for research that hasn't yet been carried out.
- **Critical abstracts:** They provide a judgment or comment about a study's validity, reliability, and completeness in addition to the main findings. As opposed to other types of abstracts, they can bring in outside information to compare and contrast. Because of their analytical commentary, they can run as long as 500 words.
- **Highlight abstracts**: This is the newest genre of abstracts. They often appear as bullet points and are written to spark the learner's attention, much like you see in the "In This Chapter" that opens each chapter of this book. There's no pretense that the abstract is a balanced or complete picture of the overall paper.

Preparing an Informative Abstract

Informative abstracts provide the learners with key aspects of the document. You write informative abstracts to stand alone or to be part of a lengthy paper. Limit abstracts to between 250-300 words or less.

Example 9-1 is the first page of a paper that appeared in a technical publication prepared for the IMAPS (International Symposium on Microelectronics) Proceedings. The brief abstract introduces the paper. The entire article was bound in a publication that was distributed at the conference. The author presented the paper at the IMAPS conference.

What to include

Following is what to include in an informative abstract:

- Subject, scope, and purpose of the study
- Methods used
- Results
- Recommendations, if any

What to omit

Following is what you omit from the abstract. Learners needing the following level of detail can read the full writing.

- A detailed discussion of the methods
- Illustrations, charts, tables, or bibliographical references
- Any information that doesn't appear in the full text

TIP

WRITING IN STYLE

In the opening sentence of the abstract, announce the subject and scope. Then you must decide what's relevant and follow with the major and minor points. Write clearly and concisely. Check out Chapter 6 to hone the tone and keep the text short and simple.

Don't forget to give yourself a byline — your name under the title. Below the title, type your name and the names of anyone else who co-authored the abstract. You may list co-authors alphabetically or in the order that represents their contributions.

Abstract portion

Stencil Printing Holds High Promise for Wafer Bumping

Jon Roberts
Cookson Performance Solutions
Foxborough, MA 02035
Phone: 508-698-7225
E-mail: jroberts@cps.cookson.com

Abstract

Solder bumping for semiconductor wafer applications requires scaling a stencil printing process from the current 50-mil geometries downwards by an order of magnitude, while driving defect densities even lower to maintain high yield. In particular, wafer bumping moves the process to smaller area ratios, where the print covers a smaller area on the wafer, but using a thicker stencil to achieve a high print volume. The effects of aperture periphery begin to dominate the printing quality, and the paste particle size approaches the of the stencil thickness. A successful process calls for an integration of printing equipment technology, solder paste development, stencil manufacture improvements and reflow furnace advances. In this paper, we describe some of the metrics used to evaluate these components and give early results of some of the tests. This work helps to determine directions for further refinement.

As a packaging choice for IC's that's been around for decades, flip-chip with solder bumps has witnessed improvements in many of the alternative die assembly processes. Although solder bump users have made their own advances, the benefits just haven't yet outweighed the problems and costs involved with its implementation. Electroplating (Solders or gold) and evaporation/sputtering as methods of making bumps haven't kept up with the advances in automated wire bonding.

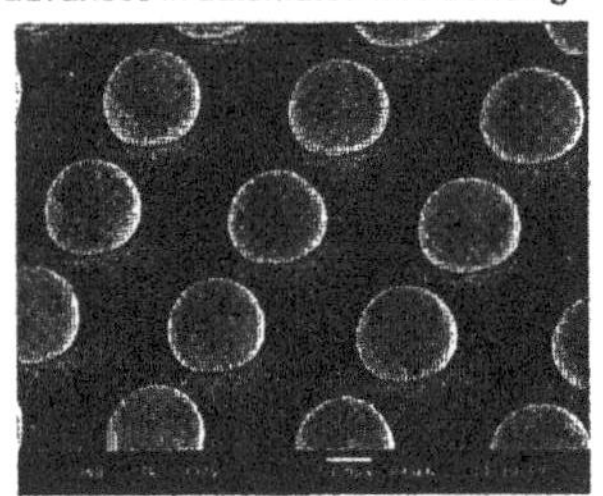

Figure 1 Typical printed solder bumps after reflow and cleaning

Stencil printing, however, brings promise of high yield and throughput, low tooling costs, and full automation to the competition. Including wafer handling and printing, a production rate in excess of 40 wafers per hour is easily achieved. Adaptable for a variety of solder paste compositions, there is no penalty for wafer size evolution and printing speed is independent of pattern density and bump size. Figure 1 shows eutectic solder bumps printed with Alpha Metals's WS 3060 paste to a pattern at 10-mil pitch, reflowed, and cleaned, ready for flip-chip or direct chip attachment.

Process integration for wafer printing calls for more than just scaling down the stencil dimensions. The typical aperture size, about 125 microns, violates the aspect ratio rule for stencils thicker than 85 microns. But high-density patterns to make large bumps require large, closely spaced apertures in thick foils. Getting past the aspect ratio rule puts increased demands on the paste release and stencil wipe processes and frequency. Reflowing the paste into uniform bumps requires the collection of all printed paste beads into the melt – flux chemistry and reflow atmosphere control are critical to successful reflow. And reflow must leave flux residues which can be cleaned with chemistries friendly to semiconductor wafers as well as to the environment.

Stencil Development Is the Key to High Yields

The stencil represents a critical limiting factor in the quality of the printed wafer. The reflowed bump's size variation can be no better than the variation of the aperture size. Although the reflow process will re-locate a paste brick to perfect position on the UBM pad, the alignment of the aperture to the pad

EXAMPLE 9-1: Highlights in abstract form.

33rd International Symposium of Microelectronics IMAPS 2000 Proceedings September 20-22, Adapted from 2000 Boston, Massachusetts

Using Abstracts Effectively

Abstracts can stand alone or be part of the longer text. When they stand alone, always let learners know where they can find the full document — in print and/or electronic form. Following are examples of how abstracts may be used:

- Some companies distribute abstracts at trade shows, conferences, or seminars to generate interest in their products or services.
- Abstracts are often sent to management-level folks (inside or outside the company) to give them a thumbnail version of a topic.
- Abstracts appear in technical journals with information on how to find the actual document. (You get an abstract published in a journal in much the same way that you get an article published. Check out Chapter 20.)

CROSS REFERENCE

When an abstract is part of the longer text that appears in a professional journal, it precedes the article. When the abstract is part of a report, place it immediately following the title page.

REMEMBER

There's a big difference between an abstract and an executive summary. An executive summary is typically 1-3 pages and covers the information in greater detail than an abstract. It may include charts, graphs, or other visual aids that summarize the full text into a few pages. An executive summary never stands alone; it's always part of the body of a long report. Check out Chapter 13 for more information on writing an executive summary.

IN THIS CHAPTER

- » Defining a spec sheet
- » Knowing what to include on your spec sheet
- » Showing example spec sheets

Chapter 10
Creating Spec Sheets

Visualize this thing you want. See it, feel it, believe in it. Make your mental blueprint and begin.

—ROBERT COLLIER, AMERICAN SELF-HELP AUTHOR

Spec sheets (the shortened version of *specification sheets*) are often called data sheets, specs, or specs on drawings. The most common types of spec sheets are written for hardware, software, manufacturing, and safety. Spec sheets can be prepared as blueprints to outline the product a company will be building, what it's going to look like, its specific requirements and functions, intended users, safety requirements, and more.

These are analogous to construction drawings, which are the backbone of every construction project. Spec sheets can be issued to detail safety requirements (as you'll see in the examples at the end of this chapter). Spec sheets may be written by engineers, technicians, programmers, and technical writers.

CROSS REFERENCE

See Chapter 7 to check the readability of your documents to ensure that your intended learners will understand them. Remember to fill out the Technical Writing Brief which is described in detail in Chapter 3.

Knowing What to Include

Every product specification is based on technical requirements, engineering specifications, and other details specific to the particular product. Language may be complicated and the spec sheets will undoubtedly feature charts, diagrams, and data sets. Technical writers are often involved in preparing specs sheets, working hand in hand with technical groups. Here's some of what to include in a product spec sheet:

» **Scope:** Product concept, what the product will look like, what features it will have, and how long the development phase should last.

» **Business case:** Benefits and advantages of the product, budget, and needed resources.

» **User personae:** Target demographics generally derived from surveys about what customers want.

» **Functional specs:** Scale, functionality, complexity, and how users will interact with the product.

» **Packaging and shipping specs:** Type of packaging required, shipping weight, temperature requirements, and more.

TIP

Check out "data sheets online" or "spec sheets online" to find templates you may be able to use.

Following the Natural Order of Things

You write spec sheets in sequential order because one builds on the other. They're always works-in-progress and need to be updated as the project changes, just as blueprints change as a building project goes along and builders learn what does and doesn't work.

Following is the order in which spec sheets may be developed:

Phase 1: Requirement specs

Phase 2: Functional specs

Phase 3: Design specs

Phase 4: Test specs

Phase 5: End-user specs

Phase 1: Requirement specs

When a company plans to introduce or update a product, it writes requirement specs to provide the design group with a starting point. These specs may be an outgrowth of a marketing analysis that mirrors the needs of the marketplace. At the very least, requirement specs include the following information:

- **Definition of the application or product:** Detail all available information about the application or product.
- **List of functions and capabilities:** Include information about what the application or product is capable of doing.
- **Estimated cost of finished product:** Create a ballpark estimate of what the application or product should cost to make it competitive in the marketplace.

Phase 2: Functional specs

Functional specs expand on the list of capabilities iterated in the requirement specs. They deal with how the system operates. Here's some information to include in a functional spec:

- **System overview:** Explain the objectives, capabilities, methods, system operation, output, and what the system will and won't do.
- **Data dictionary:** Include all the facts and figures that enter the system or are produced by it. This includes data names, abbreviations, data sources, range for numerical data, definitions or data descriptions, and codes and their meaning.
- **Input description:** Describe data that comes into the system and how the end user enters the data. This is key to the spec because the users are responsible for input; therefore, everything is geared toward ease of use.
- **Operation description:** Explain what the system will do, under what conditions, and what will result.
- **Calculations:** Include formulas to determine how the system generates numbers for output.
- **File description:** Describe the purpose, content, use, and structure of the data files.
- **Other stuff:** Explain anything that's unique to the application or product.

Phase 3: Design specs

If you think functional specs are long, wait until you see the design specs. They add yet another level of detail. Design specs may include any or all of the following:

- **Relevant documents:** Include documents that have information relevant to the application or product. This is a critical resource to the technical writer who writes the manual.
- **Functional description:** Include a detailed description of the functionality. This description may be in the form of words or diagrams.
- **Interfaces:** Discuss interfaces to the product, including power requirements and more.
- **Programming considerations:** Deal with all aspects of the functionality the programmer works with.
- **Reliability:** Describe how reliable the product is and how often it should be serviced, maintained, or updated.
- **Diagnostic issues:** Describe the testing and evaluating required to ensure a quality product.
- **Deviations:** Describe changes that may be necessary as the project progresses.

Phase 4: Test specs

Before you bring hardware or software to market, you must test the product under various conditions. Therefore, when you write test specs, you should include the following:

- **Relevant documents:** Include any documents that relate to similar hardware or software you've already developed. These documents may be helpful to testers.
- **Product description:** Identify what is and isn't being tested.
- **Testing method:** Provide a step-by-step description of the testing procedure. This must also include the process for recording and reporting problems.
- **Precautions:** Identify any special care or issues the tester must deal with. For example, you may want to document a secondary procedure to accomplish something (called a *work-around*) in case the primary procedure doesn't work.

Phase 5: End-user specs

End-user specs are basically product information sheets that ship with the product. They give users information about running the software or operating the equipment, health, and safety, and more. For other applications, end-user spec sheets may include features, strengths, weaknesses, product characteristics, support vendors, and more.

Considering Some Examples

If you ever bought a new car or even looked for one in a showroom, chances are the sales rep handed you a spec sheet comparing different car models. Even if you didn't buy the car, the spec sheet helped you remember the details of each car, such as its make, motor, steering mechanism, and so on. You may even have used the spec sheet to compare its features with cars of other manufacturers.

The following two examples each show the single page of a lengthy spec sheet.

- **Example 10-1:** This is a textual example from a spec sheet issued by the Occupational Safety and Health Administration (OSHA). OSHA is a large regulatory agency of the U.S. Department of Labor that has federal visitorial powers to inspect and examine workplaces to ensure the safety of all workers. The overall document contained graphs, charts, and tables.
- **Example 10-2:** This is a graphic example of a spec sheet issued by the Food Safety and Inspection Service (FSIS). FSIS is an agency of the U.S. Department of Agriculture responsible for ensuring that the commercial supply of meat, poultry, and egg products is safe, wholesome, and correctly labeled. The overall document contained text, graphs, charts, and tables (many of which were creatively prepared).

SHERYL SAYS

Kudos to the technical writer of the document in Example 10-2, who could have merely prepared a graph, chart, or table, but chose to present the information creatively, clearly, and professionally.

Hazard Communication Standard: Safety Data Sheets

The Hazard Communication Standard (HCS) (29 CFR 1910.1200(g)), revised in 2012, requires that the chemical manufacturer, distributor, or importer provide Safety Data Sheets (SDSs) (formerly MSDSs or Material Safety Data Sheets) for each hazardous chemical to downstream users to communicate information on these hazards. The information contained in the SDS is largely the same as the MSDS, except now the SDSs are required to be presented in a consistent user-friendly, 16-section format. This brief provides guidance to help workers who handle hazardous chemicals to become familiar with the format and understand the contents of the SDSs.

The SDS includes information such as the properties of each chemical; the physical, health, and environmental health hazards; protective measures; and safety precautions for handling, storing, and transporting the chemical. The information contained in the SDS must be in English (although it may be in other languages as well). In addition, OSHA requires that SDS preparers provide specific minimum information as detailed in Appendix D of 29 CFR 1910.1200. The SDS preparers may also include additional information in various section(s).

Sections 1 through 8 contain general information about the chemical, identification, hazards, composition, safe handling practices, and emergency control measures (e.g., fire fighting). This information should be helpful to those that need to get the information quickly. Sections 9 through 11 and 16 contain other technical and scientific information, such as physical and chemical properties, stability and reactivity information, toxicological information, exposure control information, and other information including the date of preparation or last revision. The SDS must also state that no applicable information was found when the preparer does not find relevant information for any required element.

The SDS must also contain Sections 12 through 15, to be consistent with the UN Globally Harmonized System of Classification and Labeling of Chemicals (GHS), but OSHA will not enforce the content of these sections because they concern matters handled by other agencies.

A description of all 16 sections of the SDS, along with their contents, is presented below:

Section 1: Identification
This section identifies the chemical on the SDS as well as the recommended uses. It also provides the essential contact information of the supplier. The required information consists of: • Product identifier used on the label and any other common names or synonyms by which the substance is known. • Name, address, phone number of the manufacturer, importer, or other responsible party, and emergency phone number. • Recommended use of the chemical (e.g., a brief description of what it actually does, such as flame retardant) and any restrictions on use (including recommendations given by the supplier).

1

EXAMPLE 10-1: Spec sheet issued by OSHA that is very textual.

Image from OSHA, U.S. Department of Labor

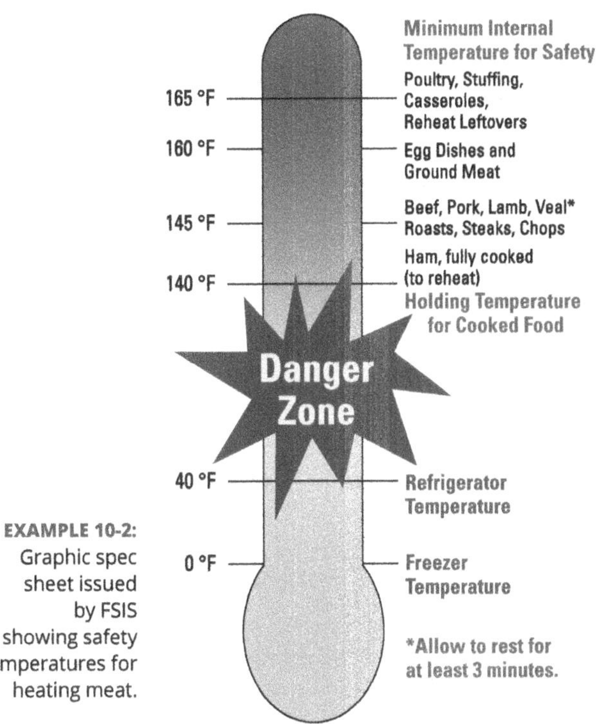

Image from FSIS, U.S. Department of Agriculture

EXAMPLE 10-2: Graphic spec sheet issued by FSIS showing safety temperatures for heating meat.

IN THIS CHAPTER

- Knowing the difference between questionnaires and surveys
- Understanding the types of questionnaires and their advantages
- Using distribution channels
- Designing the form
- Choosing the types of questions to ask

Chapter 11

Generating Questionnaires

Everything we know has its origin in questions. Questions, we might say, are the principal intellectual instruments available to human beings.

—NEIL POSTMAN, AUTHOR, EDUCATOR, MEDIA THEORIST AND CULTURAL CRITIC

I never thought much about questionnaires until I started writing the second edition of this book and decided to explore their origin. Rev. Jeremiah Milles (1714–1784) was President of the Society of Antiquarians and Dean of Exeter. As part of his research, he pioneered research questionnaires, which resulted in the "Dean of Milles' Questionnaires." The 120 numbered questions were detailed and varied. They provided rich sources of historical information about the history of parishes in Devon.

They were ultimately bound into a book that survives to this day. The book was sold at Sotheby's and is now part of the Bodleian Library in Oxford, England. Fast forward: In 1838, the Statistical Society of London drew up a questionnaire to support qualitative and quantitative data, and they're often credited with being the pioneers of questionnaires. And the rest, as they say, is history.

Differentiating between Questionnaires and Surveys

Questionnaires are any set of written inquiries to gather information from a predefined group of people. They're typically narrow in scope, length, and audience and aren't looking for a bigger picture. Questionnaires can be presented in person, on paper, by phone, or online. You can use questionnaires to:

- Prove or disprove a theory
- Determine success of a workshop, seminar, or presentation
- Measure job satisfaction
- Measure customer satisfaction
- Anticipate the success of a new product or service
- Form conclusions about a host of other ideas

Many people use the terms *questionnaires* and *surveys* interchangeably. There are subtle differences. All surveys require questionnaires, but not all questionnaires are surveys. Surveys go beyond. They're used for statistical analysis. One example of a survey is the U.S. Census, which quizzes us on everything from our ethnic backgrounds to our living arrangements, to our incomes, to our professions. The statistical analysis tells who we are and where we're going as a nation. It helps the government decide how to distribute funds to states and localities. It then helps communities determine where to build schools, supermarkets, homes, hospitals, and more.

Ask unbiased questions

While questionnaires can yield amazing insights, we must be careful because many are biased: They're targeted toward a demographic group of people to get the answers they're looking for to foster an agenda. This group may be asked leading questions to elicit certain responses. Leading questions may be:

"What problems did you have trying to assemble [product]"?

"How easy was it to assemble [product]?"

One question plants the word "problems" into the respondent's head; the other plants the word "easy" in the respondent's head. A neutral question like this is better:

"How would you describe the assembly process?"

Notice when you watch a TV program involving a courtroom, lawyers ask witnesses leading questions to put words into their months. These often result in an objection from the other attorney—but the "bias" is already out there.

Avoid double-barreled questions

There are double-barreled questions (also known as compound questions). They're two questions in one which can have two different answers. They're meant to confuse. An example would be: "How would you rate the training materials and on-boarding process?" Two separate questions would be:

> "How would you rate the training materials?"
>
> "How would you rate the on-boarding process?"

Using Distribution Channels

Distribution channels include personal contacts, postal mail, the phone, emails, and websites. Here are some things to consider:

- Not everyone has Internet access or a mobile phone. This is especially true when targeting an older population that may be less savvy about technology.
- Emails are the most popular method of distributing questionnaires and getting results. However, email questionnaires may be ignored or relegated to the "junk" files and don't reach the intended audience.
- With rising telemarketing calls, people don't often answer their phones, thereby limiting responses.
- The anonymity of online questionnaires may lead to some bizarre responses because people may use this as an opportunity to sound off, or worse.
- Questionnaires sent through the mail are still a good option for reaching a large population of people. However, there are costs involved.

In recent times, online and mobile research methods have skyrocketed. You can now conduct questionnaires and surveys for a fraction of the cost and time. This makes collecting data easier than ever before though email, QR codes, websites, and social networks. There are many tools out there to help you:

- Gather names and contact information for your targeted audience
- Provide templates

- Prepare questions with branched questions (directing the respondent to other questions based on their answers)
- Analyze trends
- Tally your results

A few popular tools (listed in alphabetic order) are Crowdsignal, Google Forms, Jotform, Sogolytics, SurveyMonkey, SurveyPlanet, Typeform, and Zoho Survey.

Designing the Form

To get the most bang for your buck, keep your form simple and brief. You want respondents to give you lots of information with minimum effort. Here are some nifty tips on preparing the questionnaire:

SHERYL SAYS

Example 11-1 is an evaluation form I give learners after attending my writing workshop. Each person has the option of leaving their name blank. Also, I use "Your Name and Title" (meaning their name); otherwise they may write in my name, as the facilitator.

Keep these points in mind as you create your questionnaires:

- **Start with an opening that lets the respondents know this questionnaire is for their benefit, not yours.**

  Able Enterprises is constantly striving to provide you with the highest quality. We value your opinion and hope you'll take a few minutes to let us know what you think of this manual.

- **Word the questions so they're straightforward, not vague.**

  Straightforward: *Have you found any content errors? If you have, please identify the error and page number.*

  Vague: *Have you found any errors?*

- **Arrange the questions so those easiest to answer (or those most important to you) come at the beginning.** Some respondents don't bother to read much beyond the first few questions, so get them to answer the important stuff while you have their attention.
- **Group together items about the same subject.** This technique makes it easy for respondents to answer the questions and easy for you to tally the responses.

» **Provide space for additional comments.** If you're doubtful whether you addressed all possible issues, give the respondents a chance to write their own comments. You'd be surprised at how much you find out.

Please let us know how we may make this manual more valuable.

We welcome additional comments. Please write on the back if you need more space.

Write It So They'll Read It™

Your Name and Title (optional) ______________________ Phone (optional) ____________

Company______________________________ Business Unit __________________

Company Address ____

I'm always looking to improve this workshop and welcome your comments. Thank you.

Workshop Content

I got the tools to	Excellent	Good	Fair	Poor
Write documents that get attention				
Help learners find key issues quickly				
Express my ideas more easily				
Karate chop through writer's block				

Instructor

	Excellent	Good	Fair	Poor
Knowledge of subject				
Confidence				
Enthusiasm				
Overall facilitation				

Productivity Results

As a result of this workshop, I expect to cut my writing time. ____ Yes ____ No
If yes, circle or add by how much? 50% 40% 30% 20% 10% ___ Other

Comments: What you found most or least valuable about the workshop. Please add any other comments.

____

____

____

____

Referrals

Most of my business comes from referrals. Please let me know of anyone who may benefit from this worshop.

Name and Title ________________________________ Phone ____________

Company______________________________ Business Unit __________________

Company Address ____

Sheryl Lindsell-Roberts
https://linkedin.com/in/sherylwrites

EXAMPLE 11-1: Evaluation form from my writing workshop.

- **Deal with confidentiality.** If the questionnaire may prove embarrassing or the results need to be private, clearly assure the respondents that results are confidential. Consider giving them the option of identifying themselves.
- **Let the respondents know how to return the questionnaire.** If you're not on hand to collect the form (such as at a seminar), provide an email address or an SASE.

Posing the Questions

A questionnaire is only as good as the questions it asks. When you don't word a question properly, answers lead to misinterpretation and skewed results. Test the wording with people who don't have a vested interest in the results. Doing so gives you honest perspectives.

There are two types of questioning techniques: closed-ended and open-ended. Some people respond well to one type and not the other, so a balance of the two often yields the best results.

Include closed-ended questions

Closed-ended questions ask respondents to select from predefined answers that are closest to their viewpoints. Questions may be yes or no, true or false, multiple choice, or a sliding scale. Closed-ended questions are easier to answer and tabulate than open-ended, but they don't allow respondents to elaborate.

Use yes, no, and in between

If you ask yes-or-no questions, give respondents a chance to opt out or respond with a middle-of-the-road answer. Following are a few examples:

Would you recommend this product to others?

- Yes
- No
- Not sure

Was the content appropriate for you?

- Yes
- Somewhat
- No

Give multiple choices

When you'd like more elaboration, consider multiple-choice questions. The following example gives respondents a number of choices:

To be responsive to the phone inquiries of our customers, how quickly do you expect answers from customer support?

- Immediately
- Within one hour
- Within two hours
- Within a day
- Other ___________

Include a sliding scale

A sliding scale gives respondents a chance to reply within a range from high to low, good to bad, or whatever is appropriate. You can use words to describe the range, as you see in Table 11-1, or you can indicate the range of numbers, as follows:

1 = Completely satisfied

5 = Not at all satisfied

Select open-ended questions

Open-ended questions let respondents answer in their own words. This approach is useful because you can get a lot of feedback from respondents' explanations. Some however, don't take the time to answer open-ended questions; that's why a mix of closed- and open-ended questions strikes a balance.

TABLE 11-1 Clearly Explaining the Range

	Excellent	Good	Average	Poor
Organization				
Appearance				
Ease of use				
Completeness				
Quality of examples				
Meets your needs				
Overall				

The following is an example of a question asking for an explanation. This provides more information than a simple *yes* or *no*. (If the answer is *no*, you certainly want to understand why in order to eliminate the problem. If the answer is *yes*, just bask in the glow.)

Would you recommend this seminar to others? Please tell us why you feel this way.

Learning from the Results

When the results of the questionnaire aren't glowing or don't give you the answers you expect (or at least hope for), learn from the responses. Always regard disapproving comments as a way to reach a higher level of excellence.

SHERYL SAYS

I offer several varieties of writing workshops and constantly tweak the content based on constructive comments from participants. My participants teach me a lot and are in part responsible for the high quality of my workshops.

TIP

Net Promoter Score (or NPS) is a widely used market research metric based on a single survey question asking respondents to rate the likelihood that they would recommend a company, product, or a service to a friend or colleague. Check out `www.netpromoter.com/know` or `https://blog.hubspot.com/service/what-is-nps`. Also check out `https://surveymonkey.com` (which is no monkey business).

IN THIS CHAPTER

- » Getting to know your audience
- » Conveying your message with confidence and competence
- » Preparing slides with just the right amount of information
- » Giving them something to remember you
- » Using a checklist to ensure you're prepared

Chapter 12

Preparing for Technical Presentations

I hate it when people use PowerPoints instead of thinking. People confront a problem by creating a presentation. I want them to engage, hash things out at the table, rather than show a bunch of slides. People who know what they're talking about don't need PowerPoint.

—STEVE JOBS, (LATE) APPLE COMPUTER GURU

Not only did the late Steve Jobs ban PowerPoint presentations, other big named-corporate giants such as Jeff Bezos (Amazon), Elon Musk (self-proclaimed Chief Twit), and others have banned them and their clones from ALL presentations. They don't call it "Death by PowerPoint" for nothing. People are bored to death viewing them.

Many people would rather undergo root canal surgery than get in front of a group and deliver a paper or give a presentation. Look at the bright side. When you give a presentation, you have an opportunity to shine and grow professionally. As a result, you can advance in your career. You may facilitate a presentation for any of the following reasons (or more):

- Representing your research and development (R&D) group by demonstrating a new practice or process
- Turning a research paper or report into a presentation
- Appealing to management for an increased budget in order to further your research
- Running a training session

Getting to Know Your Audience

Have you ever wondered why some entertainers end their shtick with, "Thank you, you've been a wonderful audience," while others are dragged off the stage with a hook? The answer is quite simple. Wonderful audiences aren't born; they aren't divine intervention, and they don't happen by accident.

Wonderful audiences are the result of the hard work of scriptwriters, comedy writers, and songwriters who take the time to understand the audience. You too can write a dynamic presentation and have a positive effect on your audience. But first you must take the time to get to know them.

TECHNICAL WRITING BRIEF

Get to know your audience by using the Technical Writing Brief in Chapter 3 and in Appendix E. Then go one step further and answer these questions about the members of your audience:

- **What do they know about me?** Will they view me as credible or must I establish my credibility? If I must establish it, I have about two minutes in which to do it. Start my talk with an accomplishment. Better yet, place my biography in an obvious place so the audience can look at it before my talk.
- **What do they know about my topic?** Do they have any preconceived ideas that make them friends or foes? For example, if I'm disputing a popular theory or medical finding, they may be adversarial. I can disarm them by saying, "I know that many of you may not immediately agree, but . . ."
- **What motivated them to attend my presentation?** They may be motivated because I'm the speaker, they have great interest in the subject matter, or their managers twisted their arms.
- **What are their objectives?** They may be there to gather information or to make personal contacts with their peers — to network.

DELIVERING A PAPER

Delivering a paper doesn't refer to the paper carrier who totes *The Wall Street Journal* to your doorstep each morning. It refers to studies, articles, or reports presented at professional conferences, conventions, or meetings. These presentations are made by expert scientists, engineers, physicians, or other technical professionals who are often asked to speak on a topic. Findings are compiled in written form and then reported at a professional conference or convention.

Make sure that your technical paper is well thought out and carefully planned. A technical paper must be presented in logical order and accomplish the following:

1. Defines the problem
2. Sets the stage
3. Explains a process
4. Shares the findings
5. Considers the broad implications

You present a paper orally, relying on notes, slides, or any other visual aids that can increase your audience's understanding of the topic. But don't be a talking head; involve the audience by asking questions and inviting comments.

Getting Ready for Prime Time

CROSS REFERENCE

When preparing for your presentation, determine five key points and sub-points that zero in on your topic. Expand each key point and sub-point as fully as you can by using the brainstorming and outlining tips discussed in Chapter 2. After you expand the points to their fullest, sequence them in the order in which they make most sense for your audience. You can find a variety of sequencing methods in Chapter 5.

Timing is everything

REMEMBER

Be aware of the time allotted for your talk. Remember that an eight-page, double-spaced manuscript takes about 15 minutes to present — without visuals. When you speak before your audience, you generally speak more quickly than you do in front of a mirror.

Get comfortable with your environment

If you're able to, check out the room beforehand. Make sure that you have all the equipment you need, including a podium, audiovisual equipment, and name tents. Find out whether you're presenting near an airport, a main highway that has trucks whizzing by, or an area where heavy construction equipment is being used. You may not be able to do anything about distracting background noises, but you must prepare for them. For example, build in a little extra time if you have to pause because of loud noise coming from construction equipment or ambient distractions.

Conveying Your Message with Confidence and Competence

CROSS REFERENCE

Use language that's clear, concise, and conversational. Keep it short and simple. Use positive words and the active voice. Be sensitive to word associations, sarcasm, and offensive, non-inclusive language, and DEI guidelines (all outlined in Chapter 6). Here are three specific ways to strike the right tone and convey your message:

1. **Phrase sentences so that they're strong and have impact.**

   Strong: Financial planners believe that the market will continue to rise.

   Weak: There's a belief among financial planners that the market will continue to rise.

2. **Project your voice to highlight the words or phrases you want to stress.** Use boldface or a marker on your notes to indicate where you want to add stress with your voice. Notice how that's done in the following sentences:

   The goldfish is in the sink. (Simple statement of fact. Nothing is stressed.)

   The ***goldfish*** is in the sink. (As opposed to the shark.)

   The goldfish ***is*** in the sink. (In case you doubted it the first time.)

   The goldfish is in the ***sink.*** (As opposed to in the bathtub.)

3. **Use statements and pauses.** Make a bold statement. Pause. Say, "Think about that for a moment." Pause again.

Use repetition strategically

You don't want to be repetitive to the point of boring your audience, but you may use repetition strategically to strengthen key ideas. Notice how the example that follows creates a lasting image in the audience's mind.

> **Strong:** Why should we adopt this policy? We should adopt it because it will give us the competitive edge. And we should adopt it because it will give us a 25 percent profit.
>
> **Weak:** Why should we adopt this policy? Because it will give us the competitive edge and a 25 percent profit.

Leave these phrases at the door

TIP

Don't start your talk with "Good morning. For those of you who don't know me, my name is [name]." Whenever I hear that, I always wonder, "What's your name for those who *do* know you?" Simply say, "Good morning. I'm [name]."

Table 12-1 shows other phrases to omit from your talk. Your audience may perceive them incorrectly (or correctly).

TABLE 12-1 Avoid These "Speaker Says" Phrases

Speaker Says . . .	Audience Perceives . . .
I'm really not prepared.	Why should I waste my time listening to you?
I don't know why I was asked to speak here today.	Am I being victimized by someone's poor judgment?
As unaccustomed as I am . . .	Thanks for sharing that. I should've stayed away.
I won't take up too much of your time.	The speaker protests too much. This is going to be a snore-fest.
I don't want to offend anyone, but . . .	Ouch! Here comes an insult.
Have you heard the one about . . .	Don't try to be a comedian.
Just give me a few more minutes.	It's already been too long.

CROSS REFERENCE

ON THE FOREIGN FRONT

It's a small world and shrinking quickly. International travel is commonplace. If you have occasion to speak before a foreign audience (even if they're foreigners visiting the United States), you must display international savvy.

Here are a few suggestions for speaking to people from foreign countries or cultures:

- Start your talk by expressing your sincere honor at being able to address the group.
- Deliver a powerful line or phrase in the audience's native language. If you don't know the native language, ask someone you trust to translate your phrase. Or check the Internet or your smartphone.
- Be aware of any current events that surround the country or culture and be sensitive to those issues.
- Cite a well-known person from your audience's country or culture. (Make sure that person is someone your audience admires.)
- If you're talking about measurements, use metric terms. The United States is one of the few places that doesn't use the metric system. Appendix C includes a listing of metric terms.
- Never (inadvertently) insult your audience with cute remarks that are social blunders. For example, if you're speaking to people from Greece, don't announce that "in America, all the diners are owned by Greeks." That's not a compliment to them; it's an insult.

Organizing for High Impact

There are various ways to organize your presentation based on the impact you want to have on your audience. When your audience will view your presentation as positive or neutral, place the key issue first. Why make them wait until the end? When you have an audience that may view your presentation as negative, build up to the key issue. Here are some tips:

- Start with a summary or a brief overview. State your key issues up front.
- Clearly state the problem/need.
- Solve the problem/need by making recommendations.
- Use transitions between your topics.

» Back up your recommendations. Second guess your audience and answer questions or objections they may have.

» Tactfully push for action. (Don't underestimate your power of persuasion.)

» Summarize your main points. Repeat the conclusions you've drawn.

Presenting Visuals

CROSS REFERENCE

Despite what Steve Jobs and others say, there are times when visuals are worth a thousand words. For more information about preparing visuals, check out Chapter 5. Here are a few tips:

» **Convey one point per visual.** The point may be what, where, how much, or any single issue you want to communicate.

» **Use fonts appropriately.** Use a 24-point font for the headline and 18-point for the text. (Your visuals must be easy to read, even from the worst seat.) Use uppercase and lowercase, even for the headlines.

» **Select colors carefully.** You don't want your audience to squint at white text on a pale green background.

» **Limit your text.** Limit each visual to between five and seven double-spaced lines of text. Use bulleted or numbered lists where appropriate.

» **Limit data to what's absolutely necessary.** Never put two graphs on one visual. If you have two graphs, use two visuals.

» **Label axes, data lines, and charts for easy understanding.** For example, the vertical data line may be sales in increments of thousands, and the horizontal data line may be calendar months or years.

» **Keep chart lines thinner and lighter than data lines.** Use lines to create structure, not to overpower the visual.

» **Use color to punctuate your message.** For example, if you use blue bars in a graph, you may want to strategically create a red bar to highlight a value you want to emphasize.

TIP

In addition to PowerPoint and Keynote (for Macs), check out `https://prezi.com` and `https://canva.com/create/presentations` to prepare charts, tables, engaging animation, and other visuals.

TIP

PRACTICE MAKES PERFECT

Practice! Practice! Practice! Even though you use notes or a fully written speech, you must sound *conversational* and make eye contact with your audience. Following are some practice tips. I don't guarantee that they'll make you a spellbinding speaker, but you certainly will make a good impression and deliver your message in the best possible way.

- Practice in front of a mirror or in front of peers who will be constructive.
- If possible, practice in the room in which you'll make your presentation.
- Practice with your visual aids, not just your notes.
- Practice without your visual aids in case "lightning strikes" and you lose electricity.
- Record your talk to hear how you sound. (If you have the capability, ask someone to videotape you to see how you appear to your audience.)

Giving Them Something to Remember You By

Always give the audience a handout or "leave-piece" — something to remember you by. This could be your business card, a relevant article, material that supports your presentation, or a paper copy of your presentation. If you leave a paper copy, consider including annotations, questions, or a place for notes. Include your name and contact information, because this is a great networking strategy.

Give handouts before your presentation

Some presenters distribute handouts before the presentation and reference them throughout. This approach is appropriate for a training session or a presentation during which you want the audience to follow along. A disadvantage may be that the audience reads the handout rather than listen to you.

Give handouts after your presentation

Some presenters save the handouts until the presentation is complete. You may want to use this approach when the handouts include data that supports your presentation. The disadvantage here is that the audience doesn't have a chance to ask questions. Also, some may run out the door without grabbing it. Others may just stuff the handouts into their briefcases as they run out the door and never read them.

Never apologize for anything in your slides or handouts. Redo anything you're not proud of. For example, while showing a slide, don't say, "This isn't current data." Instead, change the data to make it current. If you don't, you insult your audience and leave a bad impression.

Checking Out Before Checking In

Before you deliver your presentation, here are a few points to double check:

- Have I confirmed the date, time, and place of the presentation a week in advance?
- Is my objective crystal clear?
- Did I learn everything I can about my audience?
- Are my visuals informative and pleasing to the eye?
- Have I practiced my presentation in front of a mirror or before my peers?
- Have I anticipated some difficult questions? Am I prepared to answer them?

IN THIS CHAPTER

- » Realizing the importance of an executive summary
- » Understanding the anatomy of a report
- » Knowing what to include
- » Delving into executive summaries for business plans

Chapter 13
Abridging for Executive Summaries

They tried to kill us; we won, let's eat!

—ALAN KING, JEWISH COMEDIAN SUMMARIZING THE JEWISH HOLIDAY OF PASSOVER

The executive summary comes by its name very logically. It's a summary intended for executives who need a condensed version of the key elements of a lengthy, formal report in order to make timely and appropriate decisions or recommendations.

An Executive Summary Is Critical

In *Effective Communications for Engineers*, author Roy B. Hughson cited the study done by Westinghouse Electric Corporation entitled "How Managers Read Reports." The study confirms that managers read the executive summary even though they may read little else. The following is a breakdown of what managers read in a report:

- Executive Summary: 100 percent
- Introduction: 65 percent
- Body: 22 percent
- Conclusions: 55 percent
- Appendix: 15 percent

ANATOMY OF A REPORT

A report is an impartial, objective, and planned representation of facts. When you write an informal report in the form of a brief memo or email message, you generally include an introduction, body, conclusion, and recommendation.

When you write a formal report, you include a lot more. The following list shows the anatomy of a formal report, even though you may not use each component in every report. For example, your report may not have a list of figures or a glossary. This sample list simply shows the whole enchilada.

Front matter

Title page

Abstract

Table of contents

List of figures

List of tables

Preface (or foreword)

List of abbreviations and symbols

Body

- Executive summary
- Introduction
- Text (including headings)
- Conclusions
- Recommendations

References

Back matter

- Bibliography
- Appendixes
- Glossary
- Index

Summing It Up

TECHNICAL WRITING BRIEF

Be certain to fill out the Technical Writing Brief found and explained in Chapter 3. It will help you identify your executive-level audience so you can focus on what's most important to them.

Most executives often don't have the time to read the details. Yet they make decisions or recommendations about personnel, funding, policies, or other key issues based on the information they digest in one or two pages of an executive summary. Make those pages action-packed and chock-full of critical information.

Include graphics

Most executive summaries include short paragraphs and/or bullets and subheadings. However, if a graph, table, or chart will help condense critical information, include it. Use whatever is professionally necessary to get the point across clearly and succinctly.

Use an appropriate tone

CROSS REFERENCE

Following are some issues to keep in mind regarding tone and terminology. Check out Chapter 6 for a wealth of information about what is and isn't appropriate. Here are a few highlights:

- **Use technical terms cautiously.** Don't use technical terms unless you're sure that the executive reading the report is familiar with them. Not all executives have technical backgrounds, so this is where the Technical Writing Brief is invaluable.
- **Show a positive attitude.** This makes me think of a story in *The Art of Possibility* (Harvard Business School Press), written by Rosamond Stone Zander and Ben Zander. It tells of two shoe factory scouts sent to Africa to prospect business. One scout sends a telegram saying, SITUATION HOPELESS [STOP] NO ONE WEARS SHOES. The other sends a telegram saying, GLORIOUS BUSINESS OPPORTUNITY [STOP] THEY HAVE NO SHOES. If these headlines were presented in an Executive Summary as "Findings," which would please the executive? It's a no-brainer.
- **Use the active voice.** The active voice is stronger and more alive than the passive voice. For example, the active voice says, "Jose will present his findings next Friday." When you use the active voice, you place the focus on the doer of the action. When you use the passive voice, you place your focus on the action, nor the doer. The passive voice is dull and weak, as you see in the following sentence: "The findings will be presented by Jose next Friday."
- **Use gender-neutral pronouns and inclusive language.** It's critical that you show respect for everyone.

WARNING

- **Think seriously about being funny.** Incorporating humor into an executive summary (or into any part of your report for that matter) isn't appropriate. Your company isn't paying you to be a comedian, and humor is very subjective. What you think is funny may be offensive to others, especially in a business setting.

There's a big difference between an executive summary and an abstract. An abstract is a snapshot of a long report or article. It helps the readers decide whether they want to read the long text. Check out Chapter 9 for more information on writing an abstract.

Learn from a success story

I was asked by a commander of the U.S. Coast Guard to edit a 296-page report because it was "slightly" too long. ("Slightly" was a gross understatement.) The document abounded with repetitive information about the Coast Guard ports and more. I streamlined the report with charts, tables, and graphs into a final document of 28 pages. I further pared the essence of the report into a one-page executive summary. Notice the difference in the *Before* (Example 13-1) and *After* documents (Example 13-2) — both for neutral readers where I put the bottom line right up front.

- **Example 13-1:** The *Before* document is ho-hum. Although it was condensed into a single page, nothing stands out. The actual findings, which are the most important elements, are buried in the opening paragraph.
- **Example 13-2:** Look at the *After* document. It's a quick read and executive-level readers can find the information they need at a glance. Here's why:
  - Findings — the critical piece of information — are called out at the beginning with the headline "Findings: [followed by the actual findings]."
  - An easy-to-read table calls out what the findings represent.
  - The approach is numbered, not bulleted. (It was a step-by-step approach.)
  - The background appears last because executive-level people reading this summary know the background. It needed to be included for others who may read it at a later date and not be familiar with the study's rationale.

EXECUTIVE SUMMARY

BACKGROUND

This study documents the costs and benefits of potential U.S. Coast Guard Vessel Traffic Services (VTS) in selecting U.S. deep draft ports in the Atlantic, Gulf, and Pacific coasts. The concept of VTS has gained international acceptance by governments and maritime industries as a means of advancing safety in rapidly expanding ports and waterways. VTS communications are advisory in nature, providing timely and accurate information to mariners, thereby enhancing the potential for avoiding vessel casualties. This study builds on the experience of earlier efforts and provides the most comprehensive and quantitative analyses of VTS costs and benefits.

APPROACH

The following summarizes the steps used to gather the data:

- Defining study zones and subzones.
- Analyzing historical vessel casualties.
- Forecasting avoidable future vessel casualties in each study zone.
- Estimating the avoidable consequences in each study zone, the associated physical losses, and the dollar values of these avoidable losses.
- Estimating the costs of a state-of-the-art VTS Design for each study zone.
- Comparing the benefits and costs among the 23 zones.
- Analyzing the sensitivity of relative net benefits among study zones to a range of uncertainty in key input variables.

FINDINGS

The study was done over the period of one year. The findings show that the 23 zones can be divided into three groups in terms of their relative lifecycle net benefits. The following groupings are divided into areas of sensitivity. The net benefit is positive in New Orleans, Port Arthur, Houston/Galveston, Mobile, Los Angeles/Long Beach, and Corpus Christi. It is sensitive in New York, Tampa, Portland (Oregon), Philadelphia/Delaware Bay, Chesapeake North/Baltimore, Providence, Long Island Sound, and Puget Sound. And it's negative in Jacksonville, Wilmington, Santa Barbara, Portsmouth, Portland (Maine), San Francisco, Anchorage/Cook Inlet, and Chesapeake/South Hampton Roads.

EXAMPLE 13-1: The Before executive summary, where nothing pops out.

Executive Summary

This study documents the costs and benefits of potential U.S. Coast Guard Vessel Traffic Services (VTS) in selecting U.S. deep draft ports on the Atlantic, Gulf, and Pacific coasts.

Findings

The study indicates that the 23 study zones can be divided into three groups in terms of their relative life cycle net benefits. The following groupings are divided into areas of sensitivity:

Net benefit	Port
Positive	New Orleans, Port Arthur, Houston/Galveston, Mobile, Los Angeles/ Long Beach, Corpus Christi
Sensitive	New York, Tampa, Portland (Oregon), Philadelphia/Delaware Bay, Chesapeake North/Baltimore, Providence, Long Island Sound, Puget Sound
Negative	Jacksonville, Wilmington, Santa Barbara, Portsmouth, Portland (Maine), San Francisco, Anchorage/Cook Inlet, Chesapeake South/Hampton Roads

Approach

The following summarizes the seven steps used to gather the data:

1. Defining study zones and subzones.
2. Analyzing historical vessel casualties.
3. Forecasting avoidable future vessel casualties in each study zone.
4. Estimating the avoidable consequences in each study zone, the associated physical losses, and the dollar values of these avoidable losses.
5. Estimating the costs of a state-of-the-art Candidate VTS Design for each study zone.
6. Comparing the benefits and costs among the 23 zones.
7. Analyzing the sensitivity of relative net benefits among study zones to a range of uncertainty in key input variables.

Background

The concept of VTS has gained international acceptance by governments and maritime industries as a means of advancing safety in rapidly expanding ports and waterways. VTS communications are advisory in nature, providing timely and accurate information to mariners, thereby enhancing the potential for avoiding vessel casualties. This study builds on the experience of earlier efforts and provides the most comprehensive and quantitative analyses of VTS costs and benefits.

© John Wiley & Sons

EXAMPLE 13-2: The After executive summary, where key information pops out at a glance.

Delving into the Executive Summary for a Business Plan

The executive summary for a business plan is similar to a general executive summary; however, it has a single purpose: to entice potential investors. People often write business plans to fund new business ventures or expand existing businesses.

Include the following elements in this recommended order:

1. **Opening statement:** Position your business venture to spark the interest of potential investors. This is like an elevator pitch — a mission statement — to entice investors to participate.
2. **Company description:** Include the key points of your business model, what products or services you'll provide (or are providing), and a brief description of management.
3. **Market analysis and sales strategy:** Briefly describe your research that shows you will be successful in your target market. Include competitors, demand for your product, and how your product or service will stand out.
4. **Products and services:** If you're already in business, discuss current sales, growth, and future marketing plans.
5. **Financial information:** Include the highlights of your financial situation, including current sales, profits, and investments.
6. **Future planning:** Discuss how capital investment will be used to make your goals and plans happen. If you have other investors, mention them (if that information isn't confidential).

TIP

If you're having trouble getting started, take the opening sentence or two from each of the business plan sections already written. Search "business plan executive summary example" to find many helpful examples.

Put yourself in your readers' place and read your executive summary from the point of view of an investor. Does it generate interest? Is it exciting? If not, you still have work to do.

4 Tech Tools

IN THIS PART . . .

Understand the basics of collaboration, differentiate among providers, practice proper etiquette, and learn about cloud storage.

Become acquainted with videoconferencing selections, create a culture of inclusivity, get ready for your close-up, practice proper etiquette, and zoom into the metaverse.

Grasp the elements of eLearning and different delivery methods, prepare learning objectives, create modules and chunks, address the business needs, design a storyboard, and learn about testing and evaluating.

Learn advanced ways to surf the Internet, avoid predators, decode error messages, boost search engine optimization (SEO), and know when to call in the experts.

Protect your intellectual property (IP), understand how to apply for a patent, know when to trademark and apply for a copyright, and learn how to milk the cash cow.

IN THIS CHAPTER

» Collaborating with people

» Differentiating among providers

» Learning the basics

» Practicing good etiquette

» Storing data in the cloud

Chapter 14 Collaborating with Others

A strong team can take any crazy vision and turn it into reality.

—JOHN CARMACK, COMPUTER PROGRAMMER AND VIDEO GAME DEVELOPER

A little over 100 years ago one of the most popular songs on the U.S. charts was a little ditty titled, "*How Ya Gonna Keep 'em Down on the Farm (After They've Seen Paree?)*" by lyricists Joe Young and Sam M. Lewis. It was a World War I hit that became popular after the war ended. The lyrics express concern that soldiers wouldn't want to return to their families in the United States after having experienced the glamor and culture of Paris.

The concept of that song is still applicable today. Before the pandemic, it wasn't as common for employees to work remotely. That was a special arrangement to accommodate certain people in specific cases. Back then, remote workers had a bad reputation. Many managers believed they'd be too easily distracted at home, they wouldn't be as productive, they'd shirk responsibilities, and they (managers) couldn't keep an eye on them.

All that changed with the lockdown. Office workers were forced to work from home and now that they've seen *Paree*, many do not want to return to the office full time. They developed a work-life balance by working remotely (at least part of the time). This has accelerated collaboration (that was once a buzzword) into a bourgeoning industry.

Collaboration Is about People

Regardless of how many collaboration platforms are developed and how sophisticated they may be, successful collaboration will always be about the people. Assembling the right team members. Defining rolls. Creating a shared vision. Capitalizing on individual strengths. Encouraging relationship building. Creating team metrics. Fostering innovation. Defining goals and milestones. Prioritizing communication. Making changes when necessary.

Build trust

Teams can't reach their highest levels of productivity without high levels of trust; they just can't. Without trust there's less innovation, collaboration, creative thinking, and productivity. People spend their time protecting themselves and their interests.

Trust enables cooperation, encourages information sharing, and increases openness and mutual acceptance. (And trust goes beyond the team. Trust reduces company turnover, improves morale, and decreases workplace anxiety. This translates into a healthier corporate environment, less turnover, and corporate stability.)

Here are some ways to build trust across teams so that each person feels like a valued asset:

- Model trusting behavior from the top down.
- Establish the common purpose.
- Foster open communication.
- Start by assuming everyone has good intentions.
- Create kick-off time to create social capital and build from there.
- Anticipate and address stress points.
- Be transparent.
- Show appreciation.
- Encourage collaboration.
- Get to know individual team members.

- Discourage cliques.
- Respond quickly to requests and needs.

TIP

Ernest Hemingway said, "The best way to find out if you can trust somebody is to trust them." Build opportunities to create trusting relationships, such as team events, channels for shared interests or memes, polls, theme day, product-naming contests, and other getting-to-know-you activities.

Give constructive feedback

When people work together, honest mistakes and disappointments will happen. It's easy to point fingers and cast blame. This creates an unpleasant atmosphere, lowers morale, undermines trust, and is counterproductive. Feedback must be offered constructively, not as a personal attack. *Focus on the action, not the person.*

Regardless of your role as part of a team, you'll most likely be in a position to give feedback. Positive feedback should never be overlooked because we all need it. It motivates us, builds confidence, and shows we're valued. Instead of just saying "great job" or "nice work," give a meaningful compliment that shows that you really took the time to observe their work and that you truly appreciate their contribution. Make it personal. Comment on the thought process, the well-crafted message, or whatever was truly well done. All team members should be forthcoming in giving praise, recognition, and encouragement.

When feedback is negative, on the other hand, be certain to distinguish the person from the action. For example, assume you notice several errors in a document. You may tell the writer of the document that going forward, they should take extra time to proofread and edit their work (the action), rather than attacking them personally by saying they lack attention to detail or they're careless. With a personal attack, they will be more likely to shut down and lose trust in you, rather than to listen to what you have to say now or in the future.

TIP

Give negative feedback face-to-face, rather than via email. Email is one-way communication and open to misinterpretation. If face-to-face isn't possible, phone or videoconference. These offer two-way communication where vocal tone and emotional inflection are apparent. Your feedback should be timely. Don't let days or weeks pass by. You want the work to be fresh in both their mind and yours, so that the conversation will be relevant and actionable.

Find solutions to challenges

Team members bring diverse skills, strengths, backgrounds, and perspectives. Each can benefit the project. However, it's no easy feat to bring people from different backgrounds (and different time zones) together to work towards a common goal. Depending on the combination of team members and their individual characteristics (communication skills, interpersonal skills, motivations, and so on), collaboration scenarios can either provide benefits or create obstacles and challenges that may be detrimental to the overall success of a project. Dahr Mann, American entrepreneur and film producer, said, "Trust takes years to build, seconds to break, and forever to repair." Table 14-1 looks at typical challenges and explains how to create positive solutions.

TABLE 14-1 Identify and Attack Challenges Head On

Challenge	Solution
Lack of a unified mission. Unless team members are given clear objectives and key performance indicators, businesses end up with a collection of competing silos of individuals, rather than a cohesive team.	Team leaders must provide a compelling reason to be part of the mission so everyone becomes passionate about their contributions and goals. Draw from the strengths of each team member and make them feel valued.
Feelings of isolation. As people spend less face-to-face time together, it's more difficult to build relations and trust. Comments, actions, and inactions, are more likely to be misinterpreted.	Engage team members in pre-planned programs. These can include virtual team challenges, lunch and learn, happy hour, virtual break-out rooms, lectures, gamification, and anything else that will build connections.
Distribution of roles and ambiguity. Some team members may presume that responsibility for certain roles lies elsewhere or that tasks aren't distributed equally. This can lead to conflict and decreased productivity.	This can be mitigated when the team leader clearly communicates roles from the start by defining tasks, deadlines, and expectations. The leader must also track progress and provide feedback so everyone is pulling their weight.
Accountability. When responsibility shifts from an individual to a team, lines of contributions often blur. This may lead to feeling one's work isn't recognized or others are slacking and hiding behind the team.	At the outset, the team leader must communicate the team's visions and objectives. Each person's contribution must be transparent so team members hold themselves and each other accountable.
Too much brainstorming. While brainstorming holds a great capacity for problem-solving by incorporating different skill sets and perspectives, the team may become bogged down by time-consuming brainstorming meetings and debates.	This inefficiency can be reduced by narrowing the focus of each session and attempting to balance meetings and active work.
Group think. Teams reach decisions by consensus of multiple perspectives. Problems arise when team members are rebuffed or ignored and others are swayed by authoritative colleagues. This can erode the team's chemistry.	A mechanism must be put in place to create an environment where all voices are heard and respected. Reach out to remote workers who are less visible. Listen openly. Recognize often.
Potential conflict. There's a fine line between personal disagreements and conflicts. If the former turns into the latter, it can cause unhealthy competition.	The team leader must create a safe environment where opinions can be shared freely and constructively. And any conflicts must be addressed quickly before they escalate.

REMEMBER

A tool is a tool is a tool — no more, no less. Effective project leaders recognize that it's the people using the tools who are the heart and soul of every project. Leaders should look at previous projects to learn what impediments the team faced: misunderstandings, over- or under-communicating, interpreting, technology, and such, and whether the team overcame these issues.

Collaborative Team Etiquette: Etta Kitt Says

Back in the day, there were books and columns written about politeness and manners in all sorts of situations. But Emily Post isn't part of the digital society, so we have Etta Kitt to help us with collaborative team etiquette.

Get started

Just like any social situation, virtual collaboration comes with its own form of etiquette (also called *netiquette*). Here are some things to think about when getting started:

- Define the scope or overall goal of the collaborative effort.
- Identify a team leader, team, and make sure everyone has access to the same online collaboration tools from their accounts.
- Have users customize their account profiles so they include and display their full names, job titles, profile photos, gender preference, and any other important details for easy identification by all team members.
- Get everyone up to speed ASAP and be patient with people who need extra time.
- Set team deadlines and milestones.
- Assign tasks only when the context and descriptions are complete.
- Set a virtual meeting schedule as a new project or task gets underway.
- Define security guidelines as to what can and can't be discussed or distributed outside the collaborative team.

Communicate effectively

A big difference between communicating in the digital and non-digital worlds is this: One faux pas in the digital world can stick around to haunt you for much longer. Here are some suggestions:

- » Agree on a method for team members to share new ideas, comments, and constructive feedback in a way that allows everyone to be heard but that does not hinder progress.
- » All team members and collaborators should be respectful, invited to participate and share, and feel comfortable doing so.
- » Don't waste people's time with a meeting if a phone call or email will do.
- » Instruct everyone to stay on topic when interacting, using any real-time collaboration tools.
- » Have each user turn on the appropriate notifications related to the collaboration tools so everyone is alerted when a new group message is posted, a new file is uploaded, or content is edited.
- » Take advantage of a group-scheduling or time-management tool when planning and coordinating schedules and virtual meeting times.
- » Have the team leader regularly post progress reports (daily, weekly, or monthly depending on the nature of the project).
- » When it's necessary to tweak or reevaluate a goal, deadline, or objective, everyone involved must be notified.

Using Collaboration Tools

Collaboration tools help *hybrid teams* (teams working across the room, in different cities, or across the world) work together in an efficient, productive way to communicate, organize, and brainstorm. They can range from a document where team members add creative work to workflow software that keeps everyone up to speed on tasks, subtasks, and deadlines. When they're executed in a way that works for your teams, there are many benefits:

- » **Save time:** When teams collaborate with each other, they save the organization time by achieving the end goal more quickly. Information is readily available and team members can collaborate in real time, no matter where they are.

- **Strengthen team relationships**: Collaboration helps team members feel more comfortable working together, and this reduces the challenge of hybrid workers feeling disconnected.
- **Improve project management**: When people work together and have the resources they need, they can handle setbacks collectively and quickly. (Hopefully.)
- **Enhance communication.** Real-time collaboration helps to improve workflow. No more trying to find missing emails or documents. No uncertainty over what was said, done, or already shared.

Allow team members time to train

Give team members who haven't used the platform sufficient time to train. Keep it simple. Introduce them to what they need, when they need to know. Be aware of team members who may grasp the technology more slowly. Exercise patience and give them extra training.

If you need to purchase collaborative tools, different projects require and use different tools. Classify what's needed and the respective workflows. Then match the tools to these workflows. Although these decisions are made at high levels, team members usually know this best. Value everyone's input. Otherwise you risk over-engineering and becoming a slave to the tools.

CROSS REFERENCE

Before you start any collaborative project, check out Chapter 2 to learn how a team creates a plan for the best shot at success. Pay special attention to Example 2-1, the "Who's Doing What Checklist," which (ideally) the team should fill out as a joint effort. The team may include all of any of the following: project manager, subject matter expert (SME), technical writer, UX writer, graphic designer, UX designer, and others with needed skills. These folks can be employees, freelancers (who may undertake several projects at a time), and/or vendors (who typically devote themselves to one project at a time).

Consider these prolific providers

Collaboration software is one of the most crowded technology categories. If you've used collaboration and videoconferencing platforms, you've undoubtedly noticed some cross-pollinating between the two. However, none can be robust in both platforms.

I've worked with several collaboration apps but certainly haven't tried them all. For that reason I defer to three credible sources who have tested them extensively. Table 14-2 lists each of their top five picks in recommended order (with their best picks at the top). Although not shown in the table, I'd be remiss not to highlight three key players: Google Docs, Microsoft Office, and Dropbox Paper.

TABLE 14-2 Top Collaboration Picks

PC Magazine	*Capterra*	*TechRadar*
Todolist	Quickbase	Asana
Asana	Bulldertrend	Trello
Miro	Wrike	Podio
Slack	ClickUp	Ryver
Teamwork	Monday.com	Flock

Get what you pay for

This is what I call the Goldilocks theory, which is discussed in Chapter 15. *Don't pay too much. Don't pay too little. Pay for just what you need.* Some of the products mentioned in Table 14-2 have free versions that many people use successfully. Although they may be "free," consider the total cost of ownership (TCO). For example, there may be licensing fees or costs associated with setup or systems maintenance. Also, "free" isn't a sustainable business model. Many companies have been notorious for giving away their software and are out of business shortly thereafter. Be sure to deal with a company that has a successful track record.

If you want to do your own search for "collaboration tools" or "collaboration software," you'll find pages and pages of listings. If you see the letters **Ad** (short for advertisement) before the URL, that indicates the site is sponsored and will be biased. Also, when you're evaluating collaboration tools, examine privacy and security certifications that are often found on the website's "trust" page.

Know what you need

There are many nice-to-have features, but here are several should-have features.

- Intuitive interface
- Cloud-based file storage
- Cross-functional collaboration

- Internal messaging and external notifications
- Shared calendar views
- Security features
- Mobile-friendly design
- Accessibility

Be wary of free tools that solve a single problem. It's easy to download a free tool, set up an account, then download another free tool to solve another problem. Free tools can proliferate until you have a hodgepodge of tools that lead to collaboration chaos.

Perform an accessibility check

Compliance with the American with Disabilities Act (ADA) isn't an option. It's a government mandate and needs to be a business priority. Many apps come with Accessibility Checkers so your documents can be read by people with all levels of abilities. Here are some of the guidelines to make your document accessible:

- Include ALT text to provide an audio description of what's on the screen.
- Use commenting and suggesting features so screen readers can jump to this information using keyboard shortcuts.
- Use high-color contrast to make text and images easier to read.
- Use informative link text (preferably the title of the page) for screen readers.
- Include bulleted and numbered lists and bold headings in a large font.
- Present slides with captions.
- Publish to the web. Based on your account settings, you can restrict access to only people whom you designate.

An ounce of prevention. . .

When you do your homework up front and plan properly, you're on track for a smooth implementation. However, Murphy's Law may create some roadblocks. Anticipate them to minimize the impact. These are a few of the most common possible stumbling blocks:

- **Delayed implementation.** With any new implementation, you want it up and running as quickly as possible. It's not unusual for deployment to be off schedule. Build in extra time in case it's needed.

- **Low productivity for untrained users.** You can mitigate downtime when you provide adequate training and informative tutorials for all users.
- **Absence of file sharing security policy**. Have in place and communicate a security policy before implementation so you maintain a high degree of security. Check whether an app you're considering benefits you with SSL (secure sockets layer) encryption and allows you to track and manage how internal tools are accessing your data.
- **Delayed responses**. When collaborating in person, you can typically get answers immediately by walking over to a co-workers' desk. With online collaboration, you're dependent on others to respond in a timely manner. As a team, negotiate strategies for dealing with urgent questions from multiple remote workspaces and time zones.
- **Be wary of pricing (when applicable).** Choose an app that offers a fair pricing structure. Some apps are free; others will force you to buy licenses you don't need. Ask lots of questions and make sure to get answers.
- **Poor project management.** Plan the outcomes before inviting colleagues to join a workspace and appoint effective managers to oversee progress. (This is true even when everyone works in the office.)
- **Important documents are overwritten.** Agree on filenames and which people have access to what information.
- **Client collaboration and file sharing is not clearly defined.** You want only designated people giving input. The more people who share the files, the more difficult the project becomes to manage.
- **Feeling out of the loop.** Managers don't put much thought into the casual interactions that happen among co-workers in a shared physical space. Therefore, good managers must foster connections among hybrid co-workers by investing time in team building.

Storing Data in the Cloud

Most popular software today is cloud-based, either private, public, or hybrid. The cloud refers to servers that are accessed over the Internet and the software and databases that run on those servers. Example 14-1 shows a visual of the cloud system.

EXAMPLE 14-1: Sending information to the cloud.

Syda Productions/Adobe Stock

Here are some advantages of the cloud:

- » Documents are stored off-premises and typically outside the company's firewall, making the data safer than stored on individual computers.
- » This approach mitigates the risk of losing information that's saved locally.
- » On-demand access is from anywhere on any device.
- » With everyone accessing the same information, you maintain consistency in data, avoid human error, and have a clear record of any revisions or updates.
- » There is a quicker response to security updates and improvements.
- » Clouds minimize the cost of upgrades and implementations.
- » Disaster recovery is much improved, so there's limited downtime in services that could otherwise lead to lost productivity, revenue, and brand reputation.

REMEMBER

Cloud storage isn't free. Users pay a fee for their cloud storage per-consumption or a monthly rate. There may also be fees for over- or under-provisioning (reserving capacity), non-cloud services, support costs to track down issues, exit fees, or hidden cloud data transfer costs.

Understand governance

The Cloud Security Alliance (CSA) is the world's leading organization dedicated to defining and raising awareness of best practices to help ensure a cloud-secure computing environment. There are many companies specializing in securing cloud data. Do a search for "cloud security" or "cloud governance" and you'll find many of them, including guidebooks and checklists on how to make your data more secure.

Prevent security breaches

Despite everyone's best efforts, cloud-based storage has been breached (although it's rare). Facebook, LinkedIn, Cognyte, Marriott International, Alibaba, and others have reported breaches. This has led to different solutions: hybrid clouds and multicloud apps. These allow document sharing in public and private cloud environments and on-premises data centers, allowing for maximized security. Some data is stored on dedicated, local servers and some data is stored on the cloud, depending on its use, security level, and so on. In addition to heightened security, this approach can be more agile.

A SOFTWARE ENGINEER BECOMES A POET

The following is a poem written by my colleague, a software engineer who entered the world of poetry using his brain's yin and yang. Here's one of the poems in his book *In the Cloud: Poems for a Technological Age*. It's amazing how people can cross-thread professions and talents when using both sides of their brains.

IN THE CLOUD

by Win Treese

I look up and see the clouds.

Is my data in the cloud?

I look up and see the wispy cirrus clouds.

Is my data scattered in the wisps?

I look up and see the clear blue sky.

Is my data gone?

They tell me my data is in the cloud.

They tell me my data is safe.

I look up and I do not know

what data is in the cloud.

My data in the cloud

is somewhere on the net.

Where are the clouds?

Where on the net?

I look up and watch the clouds

slowly float away.

IN THIS CHAPTER

- Using the right app and additional equipment
- Creating a culture of inclusivity
- Getting ready for your close-up and etiquette
- Zooming into the metaverse

Chapter 15
Videoconferencing

Ready for my close-up, Mr. DeMille.

—GLORIA SWANSON IN THE CLOSING SCENE OF THE 1950 MOVIE *SUNSET BLVD.* (SHE PLAYED A WASHED-UP FILM STAR WHO CRAVED THE ATTENTION SHE ONCE GOT.)

Before the pandemic, many more people worked in the office and were surrounded by team members and other people in the organization — spending more than eight hours a day together, five or more days a week. Camaraderie happened naturally. You learned (perhaps more than you'd like) about each other's personality quirks, families, and what they do outside the office. These close relationships meant that when a member of the team needed help, they don't even have to think about whom to ask. As a team, everyone collaborated to produce the best results.

Then the pandemic hit and computer screens became filled with familiar and not-so familiar faces, as you see in Example 15-1. Videoconferencing is no longer a niche market; it's the place where meetings happen and collaboration takes place.

EXAMPLE 15-1: Thumbs up for videoconferencing.

Kateryna/Adobe Stock

Using the Goldilocks Theory for Selecting

As a technical writer, it's unlikely you'll be in a position to select a videoconferencing platform; you'll use whatever your employer or client has selected. However, if you're working with a company that's looking to buy or switch, perhaps they'll ask for your input. So, it's good to have a working knowledge of what's out there.

There's no one-size-fits-all, so the Goldilocks theory typically works: *Don't pay too little. Don't pay too much. Pay what's just right* (based on your needs and budget). Here are some things to consider:

- Cost (overt and hidden)
- Video-recording capabilities
- Group and private chats
- File sharing
- Screen sharing
- Breakout rooms
- Meeting notes and recordings

» Cloud storage
» VoIP (voice over IP)
» Security and privacy

Find the best platform for your needs

SHERYL SAYS

This is my personal disclaimer. Zoom and Microsoft Teams are the leading video-conferencing platforms, and I've used them both. (App and platform are often used interchangeably.) The following is based on research, not on a personal preference.

Zoom is probably the best-known name in the industry. Even people who've been living under a rock have heard of Zoom. It's been termed the Swiss army knife of videoconferencing because it has so many functions. Zoom has become as generic for videoconferencing as Xerox has become for photocopying or Q-tips, for cotton swabs.

Zoom and Microsoft Teams are neck in neck when it comes to video resolution. Both offer robust, reliable remote meeting solutions for team collaboration and communication, chats, whiteboards, voice sharing, breakout rooms, customizable backgrounds, ability to record meetings, meeting transcripts, cloud file storage, document sharing, and accessibility.

Microsoft Teams has been described as an all-in-one app — akin to Zoom, Slack, and Google Docs rolled into one. Zoom and Microsoft Teams combine to work well together, and enterprises are choosing to standardize both.

TIP

Let's not forget other players such as Google Meet (for Google workspace users), Webex Meetings (for high video quality), Skype, Slack, GoToMeeting, to name a few. Go to their websites to learn more.

FORBES WEIGHS IN

In this star-studded world, *Forbes* magazine gives Zoom 4.9 stars (out of 5) and Microsoft Teams 3.7 stars. *Forbes Advisors* issued a report by Janette Novak (contributor) and Rob Watts (editor) stating, "Zoom is our top videoconferencing platform recommendation for several reasons. Zoom offers exceptional online videoconference quality, robust business features, extensive integrations and is widely considered the most user-friendly virtual meeting solution on the market today."

Identify additional equipment needs

In addition to the app, you might need additional equipment. One thing you'll definitely need is WiFi or Ethernet that's reliable and can withstand all the data you'll use during calls. Some of what you need is resident in your device but may not suit your needs. Consider these issues:

- **Display unit:** You can use a laptop, desktop monitor, TV screen, tablet, or smartphone.
- **Camera:** Many devices come with built-in cameras. If yours doesn't or you're not satisfied, purchase a standalone. The size of your meeting room should be the deciding factor. Your best bet is to get a PTZ (pan-tilt zoom) camera for large rooms or fixed lens cameras for smaller ones.

  For hybrid meetings with multiple people, look into an Owl. It's a videoconferencing camera designed to keep everyone in the loop. This 360-degree camera connects remote team members with on-site employees by delivering a dynamic and immersive experience, leaving technology to work hassle-free and allowing your team to have effortless, natural conversations. Check out `https://owllabs.com`.
- **Microphones and speakers:** Most external webcams come with a built-in mic. But, if yours doesn't offer the high quality you want, invest in a headset that's wireless, offers noise cancellation, and has a good mic.

Creating a Culture of Inclusivity

Running inclusive meetings isn't just about technology; it's about the authentic involvement of all participants regardless of abilities or disabilities. Most of us make a common mistake when we set up large meetings and assume that all attendees will be able to participate fully. There may be attendees who have difficulty seeing, hearing, or have lots of other issues that keep them from fully partaking.

Nobody should be made to feel embarrassed or hesitant to request a special need during a meeting. For example, a presenter may say, "As you can see here. . ." It should be comfortable for a visually impaired person to say, "I can't see that. May I please have a verbal explanation?" Many apps today are building products for users with impairments, including screen readers, magnifiers, high contrast mode, keyboard shortcuts, relay service, and more. For example, captions are a lifeline to anyone with hearing issues, whose English is a second language, or who has to turn off the audio because of a noisy background.

I'm part of a Patient Advisory Council for my healthcare consortium. We had a videoconference and the captioning was on. One of the doctors said, "It's important to keep our patients on track." What was captioned on the screen read, "It's important to keep our patients on crack." (Seeing isn't always believing.)

Know the accessibility of an app

Zoom and Microsoft Teams are considered the most accessible. Accessibility features for real-time accessibility are in the early stages, and there's still a lot to learn. However, apps are continually pushing out new updates and features. Following is a checklist of things to consider before you make a decision so there are no (or limited) barriers for people to join a videoconference:

- Has the software been tested by people with different types of abilities?
- Is there real-time automated captioning?
- Does it allow for ALS interpreters?
- Are there keyboard shortcuts?
- Does it have the platform for chats, notes, or Q&A — and are they accessible?
- Does it allow for computer-based and phone-based audio listening and speaking?
- Is there a relay service enabling real-time participation through a communication assistant?
- Are the interfaces customizable?

To find out more, search the Internet for the app of your choice and add the word "accessibility." Example: "Zoom accessibility" or "Microsoft Teams accessibility."

Videoconferencing Pros and Woes

What happens when hybrid team members are scattered around the city, state, or world? Forming and maintaining connections for teamwork isn't always easy. People need to feel connected and empowered in order to be highly productive and part of the "whole." Videoconferencing enables team members to work remotely — whether from home, a coffee shop, a car, or under a tree enjoying the fresh air. The feeling of being connected boosts productivity, saves time, reduces travel expenses, and promotes collaboration. As with everything in life, however, there are pros and woes. Videoconferencing technology is rapidly advancing, so here are the pros and woes as of this writing:

Pros

Beyond the obvious, here are some advantages of connectivity:

- **Recording facilities:** Many programs can record meetings so they can be replayed should questions arise. Team members can also replay them if they can't attend.
- **Transcription capabilities:** Some apps, such as Zoom, Microsoft Teams, and GoToMeeting offer features that transcribe the spoken word into text that can be distributed. This eliminates the need for notetaking and the errors that may occur as a result of something being misunderstood or misinterpreted. Also, meetings often result in action items and next steps, and there's a record of what those will be.
- **Translation abilities:** When multi-lingual communication is needed, apps such as Zoom, Microsoft Teams, and Skype also offer real-time translations in a wide variety of languages.
- **Content sharing:** Viewers can display files or data and can collaborate and work on shared documents together in real time or through screen sharing.

Woes

While videoconferencing comes with many perks, these are also some challenges to overcome.

- **Less personal:** Nothing can substitute for a hearty handshake or an exchange of face-to-face smiles.
- **Misinterpreted gestures:** Even though we can see people, signs, gestures and indications are difficult to detect on a screen with people appearing within a box. And the more people there are, the smaller the box.
- **Lack of personal interaction:** Meetings feel impersonal and people are often multitasking, so attention is diverted.
- **Stifled participation:** It's more challenging for all attendees to actively engage, especially when the group is large. People talk over each other and don't always pay close attention or they're emailing, playing Wordle, or texting.
- **Connectivity issues:** Home networks may have limited bandwidth and rarely have backup options.

» **Security concerns:** Many of these apps have safeguards in place that ensure that only authorized people can join private business meetings. However, large numbers of people have been affected by breaches in security. Names, passwords, and email addresses have been stolen on the dark web. Although security has tightened, that continues to be a challenge.

CROSS REFERENCE

There are viruses, ransomware, and other evils lurking in cyberspace. Check out Chapter 17 for ways to safeguard yourself and your computer.

Are You Ready for Your Close-Up?

SHERYL SAYS

The videoconference renaissance has ushered in a new sense of self-awareness. If you're anything like I am, you notice imperfections (sounds better than *flaws)* on the screen that you don't notice in the mirror. Puffy bags under the eyes. Glasses that look like solar bursts. Magnified facial wrinkles. Sagging neck.

Rather than getting Botox or wearing turtlenecks in the blazing heat of summer, here are a few tricks to make you ready for your close-up:

» **Camera:** Many laptops have built-in cameras that suck in low light. They're super-wide-angle cameras that capture more of your background than you. It may be wise to buy a standalone camera with a lens of about 30mm to 50mm in focal length that's ideal for videoconference calls.

» **Lighting:** To prevent screen glare, don't place lighting behind you because it will shine directly on the screen. The best location is either resting on or mounted just above the desk. Put it on the same side as any paperwork you may be using. Also, overhead lights and desk lamps work best with LED bulbs.

» **Glare on glasses:** Once you've dealt with the lighting, lower your chin slightly but still make eye-level contact with the camera. You can also lift the earpieces a notch to increase the angle of the lenses to the light source.

» **Dress:** Avoid stripes, herringbone patterns, animal prints, and very bright colors that will distract from your face. Choose clothes that you would wear to a face-to-face meeting.

Picking a background is something else to consider. A well-thought out background can help viewers relate to you and build rapport. Opt for a background that will best represent you to your viewers. If you're meeting with upper-level management of a large company, consider a professional background of (perhaps) a bookcase, open cabinet, or other office-like furniture. If you're meeting with a co-worker, let your background reflect who you are professionally and personally. For example, if you display your interest in sailing, you'll build rapport with other sailors. Backgrounds are easy to change, so you can pick one that works for each occasion.

Fostering Cohesive Hybrid Teams

Although face-to-face contact may be lost, help your team feel connected wherever they are. There are many team-building solutions from simple, spontaneous games to pre-planned programs. I hope you'll find inspiration in this list (which is far from inclusive). Build a plan to keep your hybrid team engaged and connected:

- Start with uplifting news.
- Create a virtual break-out room.
- Hold virtual company events such as luncheons or happy hour.
- Encourage recognition and offer appreciation.
- Engage in team challenges or games.
- Introduce learning sessions, workshops, lectures, and classes.
- Connect team members who generally don't work together.
- Organize a virtual workout session with team-building exercises.
- Offer a meditation session.
- Start a virtual book club.

Suffering Burnout or Videoconferencing Fatigue?

If you're suffering from videoconferencing fatigue, you're not alone. When we were first locked down in the early days of the pandemic, everyone gravitated to Zoom to have "lunch" and "happy hour" with family and friends. We chatted and celebrated birthdays and other events digitally. That became our only means of seeing people. For many, it became a lifeline. Although the world has opened up, many people continue to work virtually, and videoconferencing remains the preferred way to conduct hybrid meetings. As a result, many people are suffering from videoconferencing fatigue.

There are several reasons why this happens. Maintaining unwavering eye contact with someone for perhaps an hour is against everything we were taught as kids. Can you recall your mom saying, "It's not nice to stare." That's what we do. We stare into the eyes of others on the screen for long periods of time.

Videoconferences cause our brains to work overtime. It requires constant mental processing to interpret non-verbal cues, such as facial expressions and other body language. This malaise has become so commonplace that researchers are studying the psychological effects it's having on people. And it's not good. Headaches. Anxiety. Muscle aches and pains. And worse.

Recognize the signs

Here are some signs that you may be suffering from videoconferencing fatigue:

- Feeling exhausted after a meeting.
- Constantly rescheduling meetings.
- Losing focus or multitasking during meetings.
- Becoming irritable at the thought of another meeting.
- Suffering muscle pain from sitting in the chair and staring at the screen for too long.
- Shouting blasphemous words at your computer.

Determine other options

Never hold a meeting for something that could have been addressed in an email or by a phone call. Videoconferencing is being hailed as the new email, but that's not entirely true. A videoconference is two-way communication — the closest thing to meeting face to face. A phone call or conference call is also two-way communication. Email, on the other hand, is one-way communication. It takes people time respond. It lacks a personal connection. Messages are often ignored, misinterpreted, or lost in the deluge of inbox messages. Many never get responses.

Always ask yourself: What's the most appropriate method of communication and what's the intended outcome?

- Do you need the immediate feedback of two-way commination?
- Do you need a discussion that would require two-way communication?
- Do you need to see people's expressions?
- Do you need to hear inflections in people's voices?
- Are there things that may need to be clarified?
- Do you have a quick question that doesn't need an immediate response?
- Are many people involved?

REMEMBER

If you determine that a videoconference is the best approach, decide whom to invite. Are there people who don't need to be there but need a report of the outcome? When is the most convenient time for all (or most) involved? Take into account that you may be dealing with people in different U.S. time zones and different countries. Also, rather than a one-hour meeting, schedule a 55-minute meeting. Or instead of a half-hour meeting, schedule a 25-minute meeting. This gives people with back-to-back meetings time to stretch, use the restroom, and get something to sip or nibble on.

Planning for Success

Many of the same elements that apply to face-to-face meetings apply to videoconferences. Here are some strategies to help your videoconference run more smoothly:

- Create an agenda. Send it to everyone in advance. Stick to it.
- Align objectives so the meeting will be productive.
- Designate a moderator who will keep the meeting on track.
- Encourage participants to raise their (electronic or own) hands to speak. That may prevent several people from talking over each other.
- Suggest that people use the chat function.
- Have someone take notes if your app doesn't have that capability.
- Identify action items and next steps.
- Schedule a follow-up meeting, if one is needed.

Onscreen Netiquette: Etta Kitt Says

"Sorry, go ahead, I didn't mean to interrupt you" two people say after interrupting each other, only to start speaking again at the same time. Then there's a 30-second cold war silence. As in Chapter 14, Etta Kitt is here to share some tips for being savvy on screen.

Dos

To develop good videoconferencing habits, here are some suggestions to consider:

- Find a quiet space (without barking dogs or crying babies).
- Test your technology well in advance.
- Arrive a few minutes early.
- Dress as you would for a face-to-face meeting (at least from the waist up).
- Have all your materials organized and at hand.
- Use proper lighting.
- Use a professional-looking background.
- Maintain eye contact.
- Press mute when you're not talking.
- Abide by in-person etiquette norms.

Taboos

Here are some taboos to remember:

- Don't arrive late. Be considerate of everyone's time.
- Don't multitask — stay focused.
- Don't smoke, eat, or slurp your drink.
- Don't talk over someone else. Wait for your turn, use the chat function, or raise your hand (or your electronic hand).

ZOOMING INTO THE METAVERSE

The term "metaverse" is a portmanteau that combines the words *meta* and *universe.* It's virtual reality (VR) on steroids — the hottest technology in the digital world that's being hailed as Web 3.0. The term is somewhat difficult to define. In essence, it's digital space populated by representations of people, places, and things — a futuristic vision that extends simulations into the real world without all the physical trappings.

The metaverse will change our lives in ways that may currently seem unfathomable. You enter the metaverse through photorealistic avatars to interact in this brave new world. Metaverse can bring teams together, build communities, affiliations, and engagements, waaaaay beyond the capabilities of any videoconferencing we're using today. Companies such as Microsoft, Apple, Meta, NVIDIA, Decentraland, Roblox Corporation, and (of course) Amazon are investing heavily in metaverse technology.

WARNING

If you feel the need to vent or make a negative comment, double (or triple) check to make sure you've muted your mic or you've ended the session. You wouldn't want the rest of your team to hear you say, "Well, that was a complete waste of time. . . Oh, crap they can still hear me."

SHERYL SAYS

Wear appropriate clothing from your shoulders to your ankles. If you decide not to wear shoes, that probably won't matter. Between the waist and ankles may matter, though. I was on a team call and one of the new team members had to get up unexpectedly to retrieve something. He wore a nice shirt but violated the steer-clear-of no-pants-dress-code option. When he got up, everyone saw him in his undies. (At least he wore them; it could have been worse.)

IN THIS CHAPTER

» Preparing learning objectives

» Acquiring the software you need

» Creating modules and chunking

» Using gamification

» Designing a process and storyboarding

» Testing and evaluating

Chapter 16
Offering eLearning

People expect to be bored by eLearning — let's show them it doesn't have to be like that!

—CAMMY BEAN, VP OF LEARNING DESIGN FOR KINEO

Originally called distance learning, eLearning has become an umbrella term for any learning done electronically. In other words, it's anything other than face-to-face with the facilitator and learners in a single room. The genesis of eLearning dates back to 1959 at the University of Illinois. It was then that two gents, Daniel Alpert and Don Bitzer, launched PLATO (Programmed Logic for Automatic Teaching Operations), which changed the world of learning. Although it was primitive by today's standards, PLATO's influence remains ubiquitous in the world of eLearning. This technology lead to the Xerox Star, which was the first computer to feature a graphical user interface (GUI), before Steve Jobs brought GUI to Apple. And the rest is history.

Understanding the Forms of eLearning

eLearning can take several forms. We live in an "always-on" society where people expect learning to be at their fingertips. They use webinars, virtual trainings, podcasts, streaming, Virtual Reality (VR) simulations, and more. eLearning brings knowledge closer to the point of need and includes *three primary types*:

- **Asynchronous:** Learning doesn't take place in real time. It's available online whenever learners need to access it. This is also referred to as Just-in-Time (JIT) learning. A perfect example of asynch (as it's commonly called) is onboarding, where new employees can learn in small chunks, as needed and as they become familiar with the company.
- **Synchronous:** The facilitator and learners interact online simultaneously, such as in a videoconference, interactive webinar, chat-based online discussion, and lectures that are broadcast at the same time they're delivered. This form of learning has grown in popularity since the onset of COVID-19.
- **Hybrid:** This is a blending of the two, and can also allow for in-person training.

Preparing the Learning Objectives

Learning objectives (also known as learning outcomes) describe what learners are expected to have mastered by the completion of a specific training unit or session. Each outcome must be *SMART*: Specific, Measurable, Attainable, Relevant, and Timely. Determine these objectives before you prepare the training. Then build your training around the *smarties*. Here's an example of vague versus specific:

Vague: Participants will learn how to deal with irritable customers.

Specific: Participants will develop and practice four narratives that will help to ease tensions when speaking with irritable customers.

TIP

Start each objective with a verb. Knowledge verbs may include *define*, *explain*, *examine*, or *compare*. Attitude verbs may include *accept*, *observe*, *judge*, or *verify*. Skills verbs may *include touch*, *locate*, *illustrate*, or *build*.

Once you pin down the business problem(s) the training is expected to solve, create *SMART* learning objectives. Objectives may be a percentage of correct test answers or a level of competency for a task. A good learning objective may read something like the following:

Objective: Learners will be able to identify all the elements of a peanut butter and jelly sandwich and be able to assemble a sandwich successfully nine times out of ten with the assistance of a series of steps.

Here's an example from West Virginia University Institute of Technology:

Course: Dynamics of Machines

Instructor: Dr. Bernhard Bettig

At the conclusion of this training, learners should be able to:

- Distinguish kinematic and kinetic motion.
- Identify basic relationships between distance, time, velocity, and acceleration.
- Apply vector mechanics as a tool for solving kinematic problems.
- Create a schematic drawing of real-world mechanisms.

Delivering eLearning

Whether you use eLearning to keep your employees abreast of company policies, teach employees new modes of operation, or help your customers better understand your products or services, there are a number of delivery environments available.

Use an LMS

LMS stands for Learning Management System. It's a digital learning environment that manages aspects of a company's training efforts, including learner information for personalized delivery, profiles, job functions, preferences, and evaluations. LMS includes open source, commercial, installation-based, and cloud-based. The cloud-based solution allows you to scale training and delivery for a greater degree of success.

SAFEGUARDING YOUR INTELLECTUAL PROPERTY

Content is the lifeblood of every brand, and it needs to be protected. This refers to any course content — your intellectual property. You can protect all your content (regardless of the delivery mechanism) by getting it copyrighted. Check out Chapter 18 and go online to www.copyright.gov for even more information. On the flip side, if you're putting the finishing touches on your training materials and you surf the net for the perfect graphic or video, *don't* be tempted to just grab and insert. You may be infringing on someone else's copyright. If you want to use it, get written permission.

Use an SCORM

SCORM (Shareable Content Object Reference Model) provides learners with highly interactive, engaging experiences, with greater control over the time spent on courses. SCORM standardizes the way courses are created and launched. It uses popular authoring tools; therefore, a major benefit is compatibility across environments. SCORM consists of standalone modules so that each module can be used in any other course within the LMS. It's highly interactive and transferrable.

Use xAPI

xAPI has become the popular new standard for delivering online training, hailed as SCORM on steroids. It allows learners to collect data about the wide range of experiences they've had, both through online and offline training.

Elements of eLearning

Every technical writer has a unique way of using elements to design structured training that adds value to their learners. However, a few elements should be consistent:

» **Navigation:** The navigation must be obvious and easy. Ease of navigation includes the use and placement of links; arrows; icons; previous, next, and move to buttons. Learners should be able to view their progress by going between past, present, and future modules. This is where UX writers are valuable.

» **Relevant content:** Develop the content from the learners' perspective, limiting to one learning objective per module.

» **Design and visuals:** Regardless of how wonderful the content is, learners will become bored if the design isn't engaging. Use high-quality graphics and consistent fonts and colors.

» **Interactivity:** Create dialogue between the learners and the tools. Learners assimilate information through quizzes, games, and tasks, and other interactive doings.

» **Tracking progress**: The company and the learner must be able to track progress. An LMS system could be valuable. It helps to know what's working and what's not and what additional training may be needed.

THE HOT TOPIC OF GAMIFICATION

Since childhood, we've all enjoyed playing games. Games were and still are a source of fun, fascination, and in many cases — learning. Games have become more sophisticated. They're now interactive video entertainment played on every digital screen you can imagine — even on something as small as an Apple watch.

As the word implies, *gamification* makes an environment, process, or task more game-like, as you see in Example 16-1. What was once dismissed as a "fad," is proving to be one of the hottest eLearning experiences. It plays into people's competitive instincts and incorporates rewards (even though they're virtual). This is a perfect solution to employee training, recruitment, evaluation, organizational productivity, and other types of learning. The concepts are simple:

- *Points* measure a learner's achievement in relation to others.
- *Badges* tap into the need for accomplishments and rewards.
- *Levels* encourage people to progress and unleash new rewards.
- *Leaderboards* organize players by rank, social status, and influence.
- *Challenges* encourage engagement with new tasks to complete.

Research continues to show that adopting game-thinking improves engagement, assists in completing much-needed tasks, improves learning, and encourages personal development. So, if you don't want to reinvent the wheel, look for places in your current training where you can insert gaming elements.

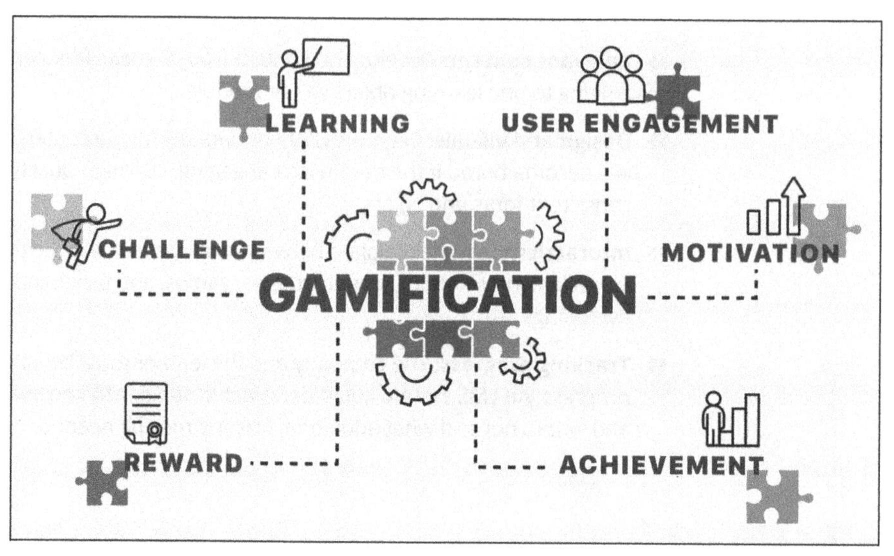

EXAMPLE 16-1: Game on!

300_librarians/Adobe Stock

Choosing Your Software

There are three ways to acquire software. Buy it. Construct it. Outsource it. In order to make your decision, be clear about your requirements, your timeframe, and your budget.

Buy it

If you need commonly used software to deal with soft skills, basic software operations, safety issues, sexual harassment prevention policies, and such, look for commercial off-the-shelf (COTS) applications. COTS is commercially produced software and/or hardware that's ready-made and available for sale, lease, or license. It's sold online and in retail stores. There are canned packages, some of which may be customized for your particular needs. COTS is the most cost-effective approach.

WARNING

Keep in mind that cheaper isn't always better. The U.S Department of Homeland Security notes that COTS software may pose a greater security risk.

Construct it

Perhaps you have in-house resources: tech writer, UX writer, instructional designer, SME, and programmer. But you don't have a videographer, narrator, or someone proficient in 3D software. Look for a vendor who offers a la carte services who will be an extension of your core capabilities.

TIP

To construct or outsource, you need to find a software development company reliable enough to give them your money and valuable time. Tap into your network. Looked at LinkedIn. Learn who built the software of your competitors. Search "software development companies." Check out `www.themanifest.com` (offering data-driven benchmarks, guidelines of expertise, and top providers).

Outsource it

Do you need a generalist or specialist? There are vendors offering a broad spectrum of services; others who specialize in certain areas such as animations or videos. The one thing you need is a vendor who will also be a partner — one who'll help you think through your requirements and add value every step of the way. Not a vendor who'll say "yes" to all your requests.

- **Consider expertise and experience:** This isn't just the number of years in business, but the number of projects they've completed that may be similar to yours. Do they have a robust website? A blog? A social media presence? And what types of information do they post? Who are their clients?
- **Get reviews:** Some vendors may not be able to show samples of their work because of non-disclosure agreements. But they should have a generic portfolio and be able to offer testimonials from clients you can speak with. You'd want to know if they were easy to work with. If their communications were open and clear. If they delivered all that was promised. If they were there afterwards as a true partner.
- **Transparent pricing:** Take time up front to understand all you're paying for. Is the pricing transparent? Are there hidden costs?
- **Trustworthiness:** This a foggy area, but an important one. Do they use licensed software? What safeguards are in place to make sure nothing is plagiarized?
- **Post-sales:** What happens after the launch? Will they be readily available for quick fixes and upgrades?

Creating Learning Modules

Learning modules are a logically structured collection of course content, much like chapters in a textbook. Learners can't possibly absorb an entire textbook in one sitting. If they tried to, their heads might explode, as you see in Example 16-2. Chapters give learners the ability to read one independent section at a time. That's what modules do for training.

EXAMPLE 16-2: Blimey! TMI.

suslo/Adobe Stock

While much training is organic and learned on the job, formal training needs to be addressed on several levels: on-boarding new-hires, compliance, products and services, software, and skills-based training. Learners need to access what they need when they need it.

- **On-boarding new hires:** New hires need to be up to speed as quickly as possible, so don't overwhelm them with too much information at one time. With many of the new hires today, games and other interactive modular activities are a must.
- **Compliance:** Compliance training is required in most companies. Employees must follow rules and regulations in the workplace. This can include corporate policies, dress codes, OSHA regulations, and health and safety issues, to name a few. Compliance training must be engaging, because most people find the topic to be boring and will put the training off for as long as possible.
- **Products and services:** Everyone needs to be aware of the products and services their company offers. This is especially true for people on the front lines, such as sales reps, customer service reps, and any others who interact with customers.
- **Software:** Although most people can use the standard word processing, spreadsheet, and presentation software, there may be specific industry-related software to learn.

ROLE OF THE TECHNICAL WRITER

The role of the technical writer in the eLearning process depends on the writer's experience and knowledge of the subject. Following are several scenarios:

- The subject-matter expert (SME) writes the draft and then turns the project over to the writer for storyboarding. (Storyboarding is explained in Example 16-3.)
- Writers who have knowledge of the subject matter write the draft and then create the storyboard and work with the SME.
- Writers with knowledge of eLearning tools and the subject matter work on the project from start to finish.

» **Skills:** Companies readily put money into training the hard skills such as computers, accounting, project management, and the like. However, soft-skill training often takes a backseat. Training in leadership, effective communication, teamwork, time management, motivation, adaptability, and such, are just as important.

Chunk information

SHERYL SAYS

On the weekends, I often cook a large pot of stew. I chunk the meat and veggies into small, mouth-sized pieces. I box them and stash them in the freezer. During the week I thaw, heat, eat, and repeat. The meals are nutritious, quick, and easy to digest. Apply "chunking" for learners — breaking information into verbally nutritious, quick, digestible bites. Chunking is particularly important with eLearning because there's no instructor to answer questions and guide the learning process.

In the 1950s Harvard psychologist George Miller conducted a study about memory that was published in an article titled "The Magical Number Seven, Plus or Minus Two." It stated that the human brain has trouble remembering more than 7 (plus or minus 2) numbers. That's why telephone numbers are seven digits long (minus the area code) and ZIP codes are five digits long.

Even numbers create symmetry, but odd numbers create interest. When people have been polled and asked to think of a number from 1-10, nearly 85 percent picked odd numbers. Therefore, it's no coincidence that the number three is persuasive throughout countless stories, fairy tales, and myths: Three Wise Men, Goldilocks and Three Bears, The Three Fates. The next time you generate a list, think in terms of three or other odd numbers. So, despite what you may have learned in the playground as a kid, being odd is a good thing.

Check for readability

When writing a training program — regardless of the delivery mechanism — arrange your content in logical chunks with the most important first. Then test the readability of your document to make sure it's at an appropriate level for your learners. Check out Chapter 7 for more information on readability.

Meet expectations

I can't say enough about how important it is to know your learners. Don't settle for a simple description of who they are. Interview some of them, if possible. You have a better chance of meeting their needs if you understand where they're coming from and what's important to them. For example, you may prepare the SMARTest content in the world, but if your navigation methods annoy them, they're not going to finish the training.

Solving Business Problems

This is an actual account of how the Technical Writing Brief came to the rescue of a tech writer. Remember that all training is designed to solve a business problem. My colleague Jennifer was assigned the task of preparing training for a janitorial staff on organic chemistry. Jennifer's client recycles a lot of the company's materials. The business problem was that the janitorial staff was having trouble keeping the *dirty* chemicals (those to be discarded) separate from the *clean* drains (where only recyclable chemicals should go). The staff members spoke a variety of languages, and Jennifer was overwhelmed at the prospect of preparing training of that magnitude.

When Jennifer filled out the Technical Writing Brief, she realized that the client had only one learning objective: Having the janitorial staff respond with the following behavior: "If you don't know what it is, don't pour it down a drain." Once she identified the learning objective, preparing the training was straightforward.

It's critical to address the following questions before writing training manuals:

- **What business problem are you trying to solve?** Perhaps support technicians in the company send too many questions up to the engineers. How can that be streamlined? Or how can you train the support technicians?
- **What change(s) do they want to make?** You may find out that the support technicians should be able to answer 90 percent of the questions on the new products without calling on the engineers. Where can they find the information?

Designing a Process

As with any good architectural concept, the vision drives the details, and the details shape the vision. Therefore, a good outcome depends on a good design. It requires knowledge of how people learn, and it deserves a lot of time and care. I remember one conversation with a gentleman who was inquiring about a career in training development. His shocking comment still rings in my ears: "I know algebra; I could write about algebra and teach algebra."

"Geez!" I thought to myself. The fact that he knows algebra has no bearing on his ability to teach the information in a meaningful way. We've all had professors who were duds. They were SMEs but couldn't communicate it clearly or effectively.

Set your sights

Envision the solution in terms of optimizing learning, the learners' expectations, the client's business requirements, and your own constraints. This includes the following:

- Time and money constraints on design, development, testing, and dissemination
- Life expectancy of the training, maintenance requirements, and updates
- Distribution channel(s)
- Level of multimedia
- Level of interactivity
- Business goals and learning objectives
- Available authoring tools and editing tools

Plan for the learner's experience

Avoid the old-fashioned model of the learner as an empty vessel to be filled with knowledge. Learning means interacting with the material so that it's meaningful. You're designing an internal experience as much as Jordan Peele designs a movie. Not quite as disturbing an experience, one would hope, but more significant for the learner. Determine the following:

- Appropriate mood to support the purpose and appeal to the learners (mood can be reflected in tone, graphics, colors, word choice, and fonts)

* Prerequisite knowledge you can expect
* Big-picture learning goals of the project
* Specific and observable learning objectives learners must reach
* Measurements your client will use to consider whether the training is successful
* Size of the average learning module
* Whether users should go through the modules in a predetermined sequence or access individual modules as needed
* How learners will navigate from one module to the next

Design the training experience

Never assume to know your learners' preferences for garnering content. Most learning is complex, cognitive, psychological, and social. It's different for everyone. About 65 percent of people are visual (or spatial) learners. They understand and remember by seeing things such as pictures and graphics. Auditory learners make up about 30 percent of the population. They retain information through hearing and listening. Kinesthetic learners are 5 percent. They retain information by doing. It may be possible to glean some of this information when you fill out the Technical Writing Brief. If you're unsure, you can't go wrong folding all three styles into the training.

WARNING

Don't succumb to including the latest tools, gimmicks, and gadgets. They're often overused, abused, and inserted into ridiculous places just because it's possible. Use them only as necessary and as appropriate.

Make a prototype

* Make a prototype of each section, representing the navigation elements.
* Test the prototype with target learners and get feedback from them as well as buy-in from stakeholders.
* Make a prototype of each section.
* Get feedback from learners and buy-in from stakeholders.

Create a storyboard

For the training developer, the storyboard is like the blueprint of a building. It outlines what goes where and the relationship between the pieces. A *storyboard* is basically a hand-drawn or computer-generated sketch of each element of the screen that marries the text and graphics, as you see in Example 16-3. (Note that this is somewhat different from the storyboard in Chapter 8. There are various ways to storyboard. Find one that works for you.)

Storyboarding isn't a glamorous phase of development, because of its repetition and trivialities, but the devil is in the details. For example, if you write a storyboard giving instructions to Zeb (the Martian you learned about in Chapter 3), you may notice that you have the image of the jelly jar on the screen, but the voice-over is talking about the peanut butter. They must sync.

If you're doing the development yourself and the modules are pretty much alike, you might get away with just one "master" storyboard. However, if you're going to delegate or if there's a significant variation between modules, create a storyboard for each module, and have the stakeholders check each module along the way.

CROSS REFERENCE

Storyboarding is very much a team effort, so be sure to fill out the "Who's Doing What Checklist" in Chapter 2 (Example 2-1) before you start the project.

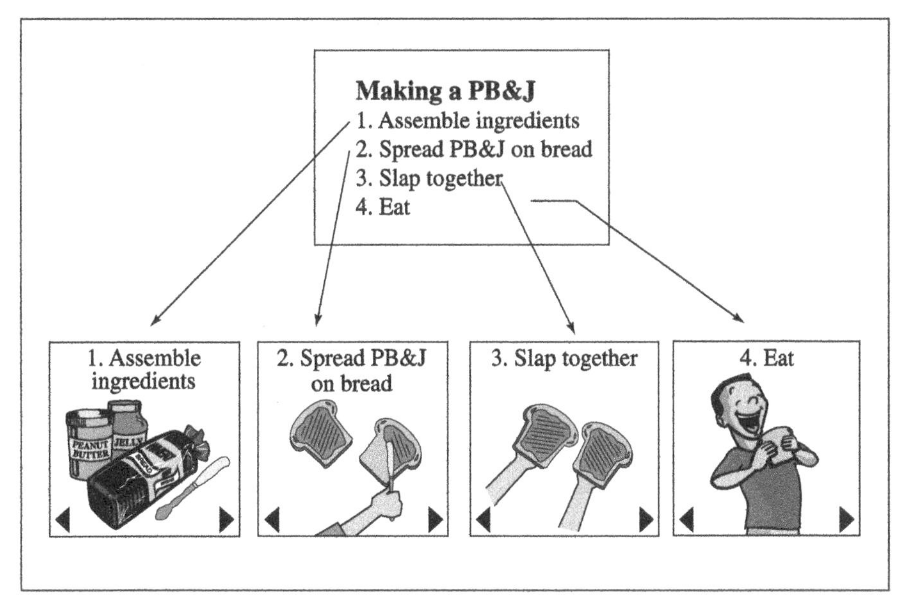

EXAMPLE 16-3: One kind of storyboard.

Testing, Testing, 1-2-3

Imagine this. You've worked diligently on a training project and release the first version. Everyone is excited and the stakes are high for you and the company. Then the big day comes. It's a disaster. Bugs appear. (Not only bugs, but large tarantulas.) This could have been avoided. Bugs and other nasties usually show up if testing is thoroughly completed.

Fix bugs and glitches

In addition to the overall functionality, you need to find and fix the little gotchas and glitches before you can say "job done." That means you (or some very patient, dedicated people) must go through every single element, choose every single menu item, press every single button, and follow every single link to be sure every single one does every single thing it's expected to do. ("Every single" is critical here.)

If you don't have the in-house capabilities of testing your training, seek outside help. Search for "software testing companies" and you'll find vendors who'll slice and dice every aspect of your training.

Evaluate training

There's a mountain of research-based evidence that links effective training to exceptional job performance to company profits. Perform periodic evaluations to assess your training needs. How do you know whether your training is effectively meeting its business goals? How do you know what's missing from existing training or what additional training may add value? How do you justify the return on investment (ROI) to shareholders, board members, and other stakeholders? Here are some suggestions:

- Prepare your own quizzes, one-to-one discussions, surveys, case studies, or exams. Your goal is to learn how employees have benefitted from training, how the company has benefitted, what content may be missing from existing courses, and what additional training may add value. Evaluations may be subjective and objective:

  Subjective: On a scale of 1-10, how would you rate [whatever]?

  Objective: Of the ten questions, how many did you answer correctly?

DON'T BUG ME

Wondering where the term "bug" originated? Back in 1878 Thomas Edison wrote a letter to his partner about an error in his machine, the quadruplex telegraph system. He referred to this malfunction a "bug." Then in 1947, Grace Hopper, a computer scientist at Harvard University in Cambridge, Massachusetts, found that the university's electro-mechanical computer, the Mark II, was delivering consistent errors. When she opened the computer's hardware, she found a moth. The trapped insect had disrupted the electronics of the computer. Even though moths aren't technically bugs (they're a paraphyletic group of insects), the name "bug" has stuck and represents any bugaboo that goes wrong with a computer.

- Determine if your LMS system has the capability of making evaluations and what the scope is.
- Hire an eLearning instructional designer.
- Go to `www.trainingindustry.com` for choices in training models.

IN THIS CHAPTER

» Watching out for predators

» Decoding error messages

» Maximizing your search efforts

» Boosting search engine optimization (SEO)

Chapter 17
Surfing the Net

The Internet gave us access to everything; but it also gave everything access to us.

—JAMES VEITCH, ENGLISH COMEDIAN

The genesis of the phrase *surfing the net* has always mystified me. I wondered what surfing the net and the sport of surfing have in common. In case you're also curious, here are two accounts:

» CERFnet is one of the first Internet service providers. (The company wanted to use the name SURFnet, but a group in the Netherlands already had it.) In October 1991, the company launched a comic book series, *The Adventures of Captain Internet and CERF Boy*. It depicted Captain Internet (a Wonder Woman-like superhero-type character) and her sidekick, CERF boy. Their adventures were titled "One If by LAN, two if by C," "The LAN that Time Forgot," "Raiders of the Lost ARP," and other computer-ese titles. (Check out `tinyurl.com/2jdtzhnu` and scroll down toward the bottom to find a photo of the comic book cover.)

» The term surfing the web is said to have been introduced by a librarian, Jean Armour Polly. (This is discussed on that website as well). Polly was using a mousepad from the Apple Library in Cupertino, CA, and she wanted to create a pithy saying to be printed on mousepads and sportswear. So she envisioned a surfer on a big wave and came up with the metaphor *surfing*. She said, "I wanted something that expressed the fun I had using the Internet, as well as hit on the skill, and yes, the endurance necessary to use it well. I also needed something that would evoke a sense of randomness, chaos, and even danger. I wanted something fishy, net-like, nautical."

Whichever account is true (both, neither, or a combo), they express going from wave to wave or from site to site. Both require patience and endurance, and can be incredibly rewarding. But both types of surfing have inherent dangers. The Internet, like the ocean, can be very dangerous if you're not cautious. Example 17-1 expresses the shared correlation.

EXAMPLE 17-1: Multitasking surfer dude.

Michael O'Keene/Adobe Stock

TIP

If you have a long URL, you can "tiny" it, as I did in the second paragraph of this chapter. Visit `https://tinyurl.com/app`. Type in the long URL, and you'll get a shortened version. For example, `www.surfertoday.com/surfing/the-woman-who-coined-the-expression-surfing-the-internet` became `tinyurl.com/2jdtzhnu`.

Avoiding the Internet Sharks

Today, the Internet is a key ingredient in every sector of our lives. It's reshaping the way we speak, think about communications, do business, and conduct research. The purpose of this chapter is to demystify the mysteries of the Internet and show how you can use it to conduct extensive research — research that may otherwise take you insurmountable amounts of time. The Internet gives access to everything from finding parts for an old piece of equipment to the latest research on just about any topic you can imagine — but you must use it wisely and safely.

As you're reading this, vicious Internet attacks are being developed. Here's how you can best protect yourself and your computer:

- **Be careful of sites that don't use https.** The "s" at the end of *http* means there's an encrypted protocol using Secure Sockets Layer (SSL), letting you know the site is secure.
- **Pay attention to the left side of the URL.** A lock sign tells you there's a secure connection. An exclamation mark tells you it's not secure.
- **Install the latest browser, anti-virus, and firewall software.** Updates typically offer the most recent protection.
- **Create strong passwords, use a different one for each site, and never divulge them.** Passwords should contain upper- and lowercase letters, numbers, and special characters. An example may be !ilove@My234!BoSS.
- **Don't let your browser save your password.** Enter your password each time you visit the site or use a secure password manager software. This is much safer.
- **Check your browser for updates.** This will give you the latest security measures, identify threats, and block unwanted popups. If you ignore these updates, your browser may become infected and the "outlaws" can get information from your search history.
- **Delete cookies.** Each website creates small bits of code to gather information from your browsing history. Intruders can steal cookies to access your accounts.
- **Use ad blocker.** This is a simple tool that removes distracting ads, making your pages easier to read. Check out "how to" on your browser.
- **Keep your virus checker running in the background.** Although this might slow your computer somewhat (most computers run at such fast speeds you'll never notice), this gives you on-going protection.

WWW = WILD WILD WEST OR WORLD WIDE WEB

When the web was born in 1989, many people chided that *www* stood for "wild wild west." That still rings true today, as there are many outlaws (even gangs of outlaws) trying to compromise devices and steal information. These crimes range from child predators to stealing information, cyberbullying, invading privacy, phishing, infecting computers with viruses or malware, password attacks, denial of service, man in the middle, emotot, and ransomware (kidnapping sites and holding them for ransom). In other words, attacking computers with all sorts of cyber assault weapons.

Cyberattacks are becoming so sophisticated, it's estimated that companies worldwide will spend $313.7 billion dollars in 2022 for cybersecurity. Although these companies may mitigate many of the problems in the corporate systems, people are working in hybrid environments, on home on computers — with at-risk sensitive information.

Dodging Other Internet Pitfalls

Search engines, like lots of things on the Internet, foster a love/hate relationship. Not all web surfers have a joy ride like you see in Example 17-1. Sometimes the surf just isn't up and it's frustrating. With all its wonders, there's lots of cursing and growling by Internet surfers. Here are some reasons why:

- **Manipulation:** Sites are ranked by paid advertisers and SEO manipulators. That may be good for the companies doing the manipulation, but they're annoying for surfers.
- **Pop-up ads:** The worst ads are the ones that take over your entire screen. Some are even streaming and mandate that you watch them in order to continue. Even blocking pop-up ads in your browser settings doesn't stop all of them (but an ad blocker can help).
- **Unrelated results:** When searching for something, weird results appear. For example, when searching for brown area rugs, you don't expect to access a site for men's hairpieces. That's probably not the area you're looking to cover.
- **Variations:** It can be confusing when you see a site ranked first on Google, and the same site is way down in Firefox because each search engine operates differently. Results are affected by how many and with what frequency web pages are analyzed by a search engine.

- **Outdated information:** Have you ever looked at a site and realized the information is no longer relevant, even though the site claims it's still pertinent? Companies don't always keep their sites updated.

Decoding Error Messages

Sometimes you get strange error messages that don't provide any useful information. Many of them have similar meanings and the conditions may be temporary. So, in most cases, wait a short while and try again. Here are a few error messages you may see and what they're trying to tell you:

- **301 Moved Permanently:** A specific web page is permanently moved to a different URL.
- **401 Unauthorized:** The page couldn't load because of invalid credentials.
- **403 Forbidden Error:** The site isn't available unless you've been preapproved. It may require a username and password.
- **404 Not Found:** The link is no longer available, in which case you may search for the information through your browser. If you don't find the site through the browser, the site may no longer be active.
- **503 Server Unavailable:** Connection to the site may have been lost or the site is temporarily down.
- **Host Unavailable:** The site could be experiencing heavy traffic or is undergoing maintenance.
- **Unable to Locate Host:** This is usually a temporary error and sometimes it's bogus.
- **Network Connection Refused:** The site may be experiencing a lot of traffic or it's only accessible by registered users.
- **The server Does Not Have a DNS Entry:** It's likely that you typed the URL incorrectly.
- **The server May Be Down or Unreachable:** The site may be overloaded, which means too many people are trying to access it at the same time. Or the site may be offline for maintenance.

Searching for the Holy Grail

Popular search engines include Google Chrome, Bing, Yahoo, Firefox, DuckDuckGo, and Safari (Mac only). The difference between a basic search and an advanced search is that you make your search more specific. Here are several advanced search methods to consider. Try a few and see what works for you.

Boolean searches

The mathematician George Boole developed a system of logic to limit the results of a search to very specific information by using the words *OR*, *AND*, and *NOT* in all capital letters. Following are some of the Boolean search methods. (With some search engines, you may have to choose Advanced Search.)

- **Einstein OR Hawkins:** When you use the word OR to separate your search components, you get a list of hyperlinks where either the name *Einstein* or the name *Hawkins* appears.
- **Science AND (zoology OR ecology):** When you use the word AND to separate your search components, you get a list of hyperlinks where the words *science* and *zoology* or *science* and *ecology* appear. The site won't contain *zoology* and *ecology* because you put "OR" in parentheses.
- **Hardware NOT (monitors OR printers):** When you use the word NOT to separate your search components, you get a list of hyperlinks where the word *hardware* appears. The words *monitor* and *printers* don't appear anywhere in the text.

Other syntax searches

Here's how to narrow your search by using other syntax:

- **Quotation marks (" ") focus your search on a particular phrase.** This treats the words inside the quotation marks as a phrase, such as "Boston Aquarium".
- **The plus sign (+) indicates a specific word or words you want in your listing.** In the following example, both mutant and secretion would appear: *+mutant +secretion*.
- **The minus sign (-) indicates specific words you don't want in your listing.** In the following example, mutant would appear but secretion wouldn't appear: *+mutant-secretion*.
- **The asterisk (*) searches for the root of a word that may have different endings.** In the following example, you would get listings for *biology, biologist,* and *biological: biolog**.

» **The percent sign (%) searches for any word that has variations in spelling.** In the following example, *criticize* is the American spelling, and *criticise* is the British spelling: *critici%e*.

Speed surfing

If your browser is slow, consider switching to a faster or newer one. Google Chrome is typically the fastest browser. If you've switched browsers and still experience issues, consider changing your Internet provider. Or perhaps your computer is running slower than usual. Try these fixes:

» **Close any tabs you're not using.**

» **Uninstall unwanted apps and programs to clear hard disk space.**

» **Defragment (defrag) your drive.**

» **Run your virus checker.**

» **Optimize your browser cache.** Each time you download web pages, they're saved on your computer. When configuring your browser caching rules, consider whether the browser should cache the site, how long it should be cached, and what would invalidate that cache. If you browse lots of sites, it can be helpful to have a large cache.

» **Clear out cookies**. If every place you visited gave you a bag of cookies to tote around, those cookies would get quite heavy and slow your walking pace. In that same way, your computer saves information each time you make a search, and these cookies (text files with small pieces of data) can slow down your computer's performance.

» **Block ads**. Many websites rely on ads to continue operating, but they're often responsible for making web pages load more slowly. Search for "ad blockers" to block those pesky ads.

Boosting Search Engine Optimization (SEO)

The Goliath of elevating sites to the top of browsers is SEO. Just because a site pops up at the top of page 1, doesn't mean that's the best search result for you. The best result for you may be on page 2 or 3. Why? Because there are many ways to force a company's site to the top, discussed next.

Understand pay-per-click (PPC)

PPC is where search engines show ads. Each time someone clicks on the ad, the company pays a fee. If no one clicks, the company pays nothing. Essentially, it's a way of "buying" sites to the top of the listings. Today, more people are using the Internet to search for products and services relevant to their needs and interests, and this is a legal way to bring traffic to companies that are willing to pay.

Go organic (non-paid results)

This is achieved by obtaining a high-ranking placement based on substantive content.

- Make your headings and subheadings meaningful by using keywords people would use to find you.
- Include bold, italics, and heading tags to highlight these keyword phrases — but don't overdo it.
- Keep your site relevant and current.
- Use metadata, especially title metadata. It's the most important metadata and is responsible for the page titles appearing at the top of the browser window.
- Identify link-worthy sites. Check regularly to make sure the links work.
- Use alternative text (ALT text) descriptions, which are a variety of ways to say the same thing. They allow search engines to locate your page. This is crucial, especially for those who use text-only browsers or screen readers. (There are alternate text generators you can find online.) The more descriptive you are, the better. Consider this example:

Before: `alt="Teacher pointing to a computer screen"`

After: `alt="College professor using education software to instruct a business school student"`

Avoid keyword stuffing

WARNING

This is considered web spam or spamdexing. It's the process of using metadata to trick search engines into making certain websites seem more relevant than others. This is considered bad form and is detrimental to your search rankings, because search engines tend to favor pages that create a great experience for users.

Turn SEO over to the experts

Unless you're a computer whiz and adept at SEO rankings, it may be wise to work with an SEO company. They're armed with loads of techniques without keyword stuffing and other tricks.

TIP

Google is the most popular search engine for PC users. For that reason, consider claiming your listing on GoogleMyBusiness. It's a free tool that allows you to take charge of the way your business appears in a search. Also check out Google Search Console to measure your site's performance, fix issues, and make your site more visible. The following tools are for Mac users:

- Ahrefs
- EverWeb
- Semrush
- SEO Auditor
- SEO PowerSuite
- Ubersuggest
- WebCEO

CONNECTING THE DOTS

Perhaps you're confused by all the *dot-whats* at the end of a web address. Although .com is the most common web extension, there are many others, and more are coming on the cyberscene. Following are extensions you commonly see and the type of sites they represent:

- .com(commercial site)
- .net (commercial site)
- .org (commercial or nonprofit site)
- .edu (educational site)
- .gov (government site)

Adding to the mix are country-specific extensions, such as the ones that follow:

- .ar (Argentina)
- .at (Austria)
- .au (Australia)
- .be (Belgium)
- .br (Brazil)
- .ca (Canada)

(continued)

(continued)

- .ch (Switzerland)
- .cn (China)
- .de (Germany)
- .dk (Denmark)
- .es (Spain)
- .fi (Finland)
- .fr (France)
- .ie (Ireland)
- .il (Israel)
- .is (Iceland)
- .it (Italy)
- .jp (Japan)
- .kr (South Korea)
- .kp (North Korea)
- .lu (Luxembourg)
- .mx (Mexico)
- .nl (Netherlands)
- .no (Norway)
- .nz (New Zealand)
- .pl (Poland)
- .pt (Portugal)
- .se (Sweden)
- .us (United States)
- .tw (Taiwan)
- .uk (United Kingdom)
- .za (South Africa)

IN THIS CHAPTER

» Understanding why protecting intellectual property is critical

» Applying for a patent

» Establishing a copyright

» Registering a trademark

» Licensing to milk the cash cow

Chapter 18

Protecting Intellectual Property

Intellectual property is the oil of the 21st century.

—MARK GETTY, CO-FOUNDER AND CHAIRMAN OF GETTY IMAGES AND SON OF J. PAUL GETTY, OIL TYCOON

You or your company came up with a great idea, a design, a training module, or a product you believe can be a real money maker and is worth protecting. It's time to apply for a patent. Establish a copyright. Register a trademark. Become a licensee. Whether you're part of a startup or a multinational company, it's important to safeguard your intellectual property (IP) from theft or copying. To begin, take these measures:

- Be careful whom you trust with your ideas.
- Ask employees, partners, and anyone else privy to your idea to sign a confidentiality or non disclosure agreement.
- Document your concepts in excruciatingly great detail.
- Apply for a patent, copyright, or registered trademark as soon as possible.

In 1875 Alexander Graham Bell was granted a patent for an apparatus designed for "transmitting vocal and other sounds telegraphically. . . causing electrical undulations." However, many years earlier, Antonio Meucci began inventing a talking telegraph (or phone). In 1849 Meucci applied to the Patent Office for a caveat. (A caveat is an official notice of intent to file a patent application at a future date — somewhat like a reservation or placeholder. The caveat application must contain all the drawings and details for filing. It's honored for one year pending a renewal.) Due to personal hardships, Meucci couldn't renew his caveat, so the phone patent was up for grabs.

Adding another twist, Elisha Gray, an engineer and professor at Oberlin College in Ohio, tried applying for a caveat to the Patent Office for a telephone that he intended to submit within three months. Learning of Gray's intent, it's claimed that Bell's attorney rushed to the Patent Office first (on the same day). There are claims that Gray had evidence of Bell's invention and tried to usurp the patent. Controversy still surrounds the invention of the telephone, but we'll never know.

So if you were a contestant on Jeopardy and needed to answer the clue: *The inventor of the telephone*, you'd ring the buzzer and shout out "Who is Alexander Graham Bell?" Your answer would be accepted as correct. But would you really be correct?

TIP

If you can, avoid joint IP ownership. It's beneficial to have sole control of your business decisions. Joint ownership may lead to a variety of disputes that can cause confusion, or worse, legal issues. Many companies have gone belly up because of joint ownership clashes.

SHERYL SAYS

This chapter touches on highlights of patents, copyrights, and trademarks as they apply to the United States. Each country has its own regulations. For more details check out `www.uspto.gov` (U.S. Patent and Trademark Office) and `www.copyright.gov` (U.S. Copyright Office).

Applying for a Patent

Anyone can apply to the U.S. Patent and Trademark Office (USPTO) for a patent regardless of age or mental competency. Even deceased people can file a patent through a personal representative. Each year, thousands of people add prestige to their names and the names of their companies and universities because of inventions. It's interesting to note that since the USPTO opened its doors in 1790, 4400 mousetraps have been patented. Who ever said, "You can't build a better mousetrap"?

If you invent something, you want to exclude others from making or selling it. Patents, therefore, are intended to promote innovation and make sure that inventors reap the financial rewards for their inventions. When you're granted a patent and someone violates it, you can sue for patent infringement. Once a patent expires, however, anyone can make, use, or sell the invention or its design. Utility patents have a 20-year term and design patents have a 15-year term. Neither can be renewed.

What exactly is an invention and what is a discovery? *ComputerUser* editor James Mathewson wrote, "We don't invent new mathematical truths, we discover them." As an outgrowth of patenting hybrid crops, biotech companies are filing patents for genetic sequences. This development has many scientists displeased. They feel that one day a person or company may own the rights to complex animals or humans.

Although filing a patent gives your exact creation protection, it can be a recipe for difficulty. The patent shows how a product or service can be created. Once it's published, others can create a similar product with workarounds that won't violate IP rights. As an example, Apple is in a "food fight" with Prepear (a cooking app) that's using the logo of an outlined green pear-with-leaf. Apple claims that Prepear's logo calls to mind its iconic bite-of-an-apple logo and is in violation of its registered trademark. Is this bullying or a violation? The courts will decide.

Know the types of patents

There are two basic types of patents — utility and design patents.

Utility patents

Utility patents cover inventions that are electrical, mechanical, or chemical. These include staplers, electronic circuits, pharmaceutical products, microwave ovens, semiconductor manufacturing processes, and much more. They even include genetically engineered bacteria for cleaning up oil spills. In order for an invention to be granted a utility patent, it must meet the following specifications:

- **Its use must be obvious.** Does the invention produce new or unexpected results?
- **It must be novel.** Is there some novel feature, or is the invention merely a new use for something already done?
- **It must fall into a statutory class.** That includes processes, machines, articles of manufacture, compositions, or new uses of another item.

Design patents

Design patents cover inventions that have a unique shape. This spans everything from computer screens, to the design of a water cooler, to telephones shaped like Elvis's guitar.

With some design patents, however . . . you just *gotta* wonder. Example 18-1 is a baby carriage invented by George Clark. It looks like a high-buttoned shoe with wheels underneath. Mr. Clark invented this carriage so that wealthy people could push their offspring around in a carriage no one else had.

EXAMPLE 18-1: Keeping kids in shoes.

© John Wiley & Sons

And who said you can't reinvent the wheel? That's exactly what Sydney Jones of Great Malvern, England, did. He invented a wheel made of elastic spring steel. Its rims fold around obstacles and roll over them, somewhat like an air cushion, as you see in Example 18-2.

REMEMBER

If two or more people work together on an invention, in whose name is the patent? If they are joint inventors, the patent is issued to them jointly. However, if one person provided all the ideas and one or more merely followed instructions, the person who conceived the ideas is listed as the sole inventor.

EXAMPLE 18-2: Reinventing the wheel.

Do your homework

The U.S. Patent Office doesn't require that you make a search to see whether your invention has already been patented; however, why put effort into an invention if someone has already claimed it? There are several ways to search:

- **Check out stores, catalogs, product directories, and more.** For example, if you invent a newfangled paper clip, check large stationery outlets and catalogs to see whether something like it is already on the market.
- **Do an Internet search for "patent search."** You find more sites than you know what to do with. For more information on conducting a search, check out Chapter 17.
- **Visit the U.S. Patent Office.** Why not enjoy the sights in and around the nation's capital and have a wonderful vacation or business trip? (Visit `www.uspto.gov`.)

Submit your idea

You don't need a finished product to apply for a patent. Before you get into the nuts and bolts (literally), prepare a detailed sketch of your idea. Example 18-3 shows a patent drawing submitted by Jon A. Roberts (my husband), Karl J. Armstrong, and Arnold J. Aronson for a planarization method used in the semiconductor industry. In addition to the sketch, here's what the inventors included as part of the packet:

- Abstract
- Background of the invention
- Summary of the invention
- Description of the drawings
- Detailed description of the invention
- Supplementary drawings

Submit ample sketches to show as many views as necessary to detail your invention. Send a computer-generated or black-and-white drawing. The Patent Office doesn't generally accept photographs for utility patent applications. If a color photo is important to displaying your idea, file a petition requesting that a color photo be accepted. If you send a photograph, develop it on double-weight photographic paper or permanently mount it on Bristol board. The photos must be of sufficient quality so that all details in the drawing can be reproduced in the printed patent.

REMEMBER

Once you've submitted your application, you can mark your product "Patent Pending" or "Pat Pend." for as long as the application pends. There's no legal benefit, but it informs the public that a patent has been filed. This may be enough to deter blatant copying. Once the patent has been approved, you can include the following:

- Patent or (Pat) 1,234,567 (for utility patents)
- Patent or (Pat) US 1,234,567 (for design patents)

Know who owns the patent

Imagine that you work for big pharma. During the course of your employment, you discover a combination of chemicals that can save hundreds of thousands of lives. You and your team develop the drug, and the company has the potential to make incalculable profit. You apply for a patent and Viola! It's awarded. But who holds the patent? The drug was your brainchild. However, the company paid your salary and provided the lab, the team, the chemicals, and the equipment for you to create your wonder drug.

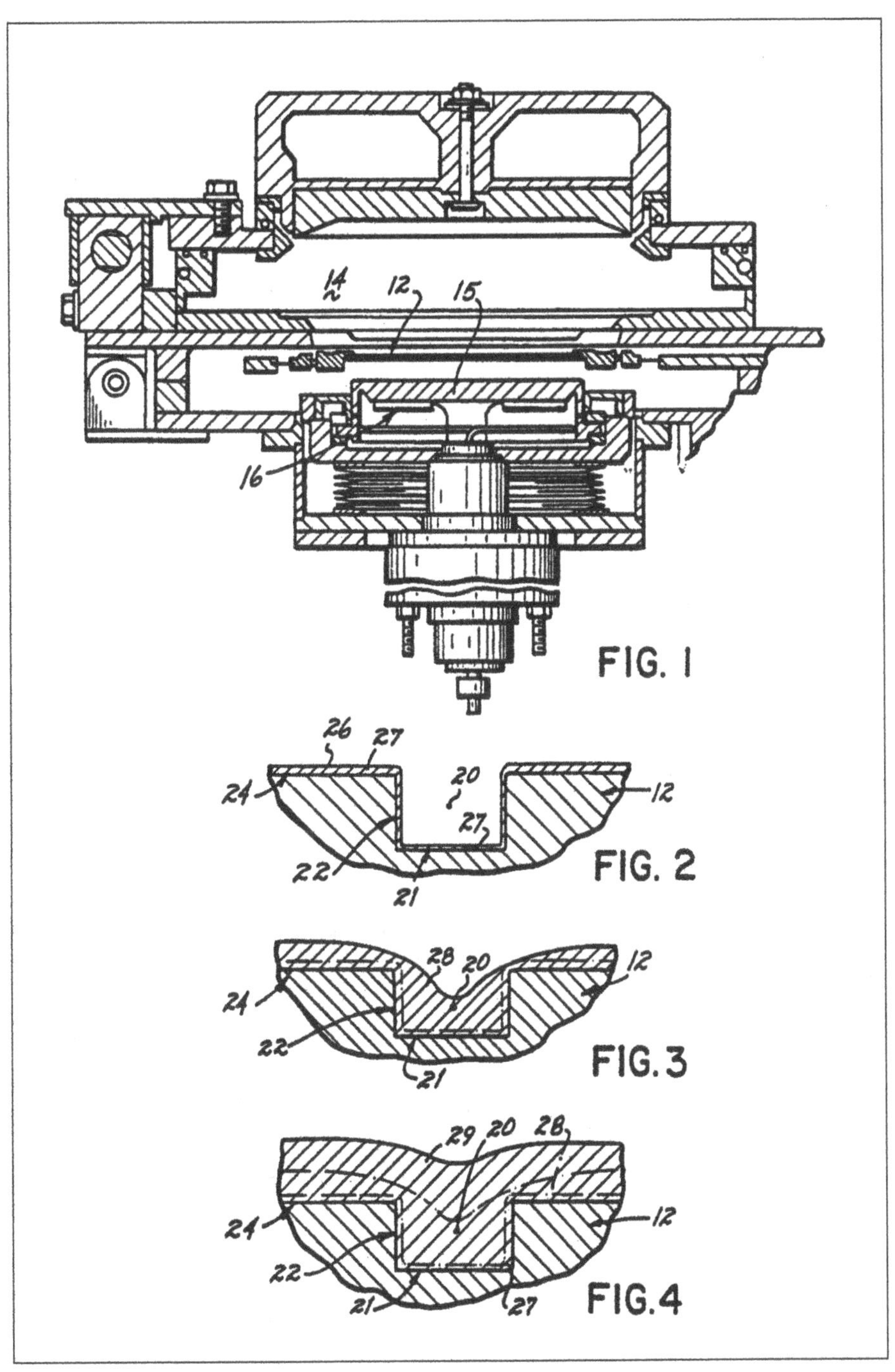

EXAMPLE 18-3: Drawing submitted for patent approval.

ON THE FOREIGN FRONT

U.S. patent laws have no effect in a foreign country. Therefore, a patent granted by the United States extends only throughout the United States and its territories. Countries have their own patent laws, and you must make an application for a patent in the country where you want your invention sold and protected.

The Paris Convention for the Protection of Industrial Property is a treaty relating to patents that 140 countries adhere to. It provides that each country guarantee to the citizens of other countries the same rights in patents it gives to its own citizens. If you contact a U.S. patent attorney, they may be able to set you on the course of acquiring a patent attorney abroad.

Most companies typically require employees and contractors to sign a written agreement assigning to them any patent that's issued during the course of the employment/contract. Fair or not, employers typically hold the patents to the creations of their employees. However, it's a huge feather in your cap and an outstanding resume "brag."

WARNING

Don't be deceived by the "poor man's patent" myth. It's a misconception that if you mail to yourself all your documentation and send it via certified or registered mail (and leave it sealed), you've patented your work. Yes, this is less costly, but it's fundamentally flawed. It's not a patent and offers no legal protection.

Establishing a Copyright

You can secure a copyright for your original work once you put it in tangible form. (Unlike a patent, you can't secure rights to an idea or concept.) Works that can be copyrighted include training materials, computer software, audio/visual productions, books, poems, dramatic works, musical compositions, sculptures, paintings, architectural works, and the like. Names, titles, short phrases, and slogans aren't copyrightable.

Copyright protection lasts for the life of the author, plus 70 years. If anyone copies or uses your materials without your permission, they've violated your rights. You're entitled to sue the copyright infringer for any profits they made from your work.

Know what to include

Present your application clearly and succinctly. Include the vital info, address the main objectives, and include features and secondary benefits. For example, if you want to copyright a new app for time management, stress how it can help professionals earn more money and work more efficiently. Less important is how it syncs up with other apps.

Get your works copyrighted

Copyrighting your work is straightforward. Go to `www.copyright.gov` and click on "Register Your Works." It walks you through the process. Here are three basic requirements:

- **Originality:** This doesn't mean novel or unique; it means what you're submitting isn't a copy of someone else's work.
- **Creativity:** According the U.S. Supreme Court, the works need to have a "modicum" of creativity.
- **Fixation:** Your works can't be purely ephemeral. As soon as you document it, however, it's considered fixed.

Fees depend on what and how you're registering. As of the time of this writing, online registration is $45; print registration is $125. Fees are subject to change, so always check.

WARNING

Don't be fooled by ads that pop up when you search for copyright information. They might look like links to the official site. To be safe, go directly to the government website at `www.copyright.gov`.

REMEMBER

You don't have to wait until you've filed your application to place the copyright symbol on your work. Some people place it without having any intention of filing so it will (hopefully) serve as a deterrent. However, it only has legal protection if you've been approved. Display the copyright notice in an obvious spot. For example, here's how you may include it on your training material:

Copyright © 20XX [OWNER'S NAME]. All Rights Reserved.

Registering a Trademark

A *trademark* can be any word, catchphrase, symbol, design, or combination that identifies goods or services. Companies live and die by their brand names. It's how customers recognize a brand and distinguish it from competitors. Unlike patents and copyrights, trademarks don't expire after a set period of time. However, to avoid losing the trademark, on the fifth and tenth anniversaries of issue, the trademark owner must file a declaration that the trademark is still in use. For more information about trademarks, check out `www.uspto.gov/trademarks/basics/what-trademark`.

Celebrities such as Beyoncé, Taylor Swift, Rihanna, Victoria Beckham, Justin Bieber, Bruce Springsteen, and others have taken the entrepreneurial route and branded their names to sell clothing, fragrances, jewelry, handbags, and other paraphernalia. However, Tom Brady, superstar for the New England Patriots, who later joined the Tampa Bay Buccaneers, was not allowed to trademark his nickname "Tom Terrific." Tom Seaver, a baseball Hall of Famer, earned that nickname many years before Brady was old enough to hit the field.

Use unregistered trademarks and service marks

A simple trademark is a mark that companies often use on a logo, name, phrase, word, or design to represent its goods or business. It's not registered and has no legal significance. It merely stakes your claim to the name, alerting competitors that you use it for business. You show a trademark for goods by using TM symbols and a trademark for services using the SM symbol:

Name^{TM} or Name (TM)

Name^{SM} or Name (SM)

TRADEMARK TRIVIA

Trademarks can become invalidated when the name becomes so commonplace that the brand is no longer associated with the trademarked product. For example, the Xerox photocopy machine was a trademarked name. It's now used generically to mean "photocopy" and is no longer associated with a particular brand of copier. This is called *genericide*. Everyday products that have been genericized include Aspirin, Band-Aids, Bubble Wrap, Chapstick, Clorox, Fiberglass, Frisbee, Lazy-Z-Boy, Memory Stick, Photoshop, Ping Pong, Realtor, Scotch Tape, Vaseline, Velcro, and so many others.

Apply for a registered trademark

There's power in the name of a business so you must ensure that the name remains one-of-a-kind. Registering a trademark helps protect your name or brand from intellectual property theft or misuse as a business expands. You can register a name as shown in a particular color or colors, style, font, or special markings.

Here are a few things to keep in mind:

- You can register your trademark with the federal government or the state in which you operate your business. The federal registered trademark gives protection across the entire country. A state registered trademark protects you from infringement only within that state.
- For a federal trademark, start by doing a Trademark Electronic search (`www.uspto.gov/trademarks/search`) to learn if someone has already registered or applied for that trademark.
- For a state trademark, check out `www.uspto.gov/trademarks/basics/state-trademark-information-links`.
- Go to `https://trademark-genius.com/registration-trademark` to start the registration process.

Once you've been approved, you can proudly display the circled R, as you see in Example 18-4.

WARNING

Never enter into a licensing agreement without consulting an attorney.

EXAMPLE 18-4: Legal registered trademark.

adragan/Adobe Stock

LICENSING TO MILK THE CASH COW

Licensing your product or service is a brilliant way to milk your cash cow — you reap a steady stream of revenue and/or deliver your expertise on a much wider scale. As a licensor (grantor of the license) it's within your purview to dictate the terms of manufacturing, marketing, and delivering your patented, copyrighted, or trademarked creation.

Licensing products: Licensing products is steeped in the big-eared, lovable character named Mickey Mouse. In 1929, around Mickey's first birthday, a man approached Walt Disney in the lobby of a New York hotel and offered Walt $300 in cash for the rights to produce Mickey on a children's pencil tablet. Needing the money, Walt readily accepted the offer. Since then, licenses have been issued to a plethora of manufacturers for Mickey to appear on watches, costumes, dolls, bed sheets, lunchboxes, backpacks, and the list goes on and on! Disney has become the top licensor in the world, reaping double-digit billions of dollars in retail sales.

Licensing software: Licenses can be issued for public domain, subscription, or perpetual-use software. For example, when you take a bite of the Apple, you've agreed to abide by Apple's licensing terms and conditions in effect at the time of purchase. Windows 11 Enterprise (for example) is licensed as an upgrade to Windows Pro. Copyright laws don't apply to works in the public domain such as SQLite, 12P, or CERN. Public domain software has no ownership and can be used and modified without restrictions.

Licensing training materials: Licensing all or part of your training materials will help recoup your development costs and create a passive income model while maintaining the accuracy and integrity of your copyrighted creation. Check out "pricing models for licensing training" to find a variety of revenue-producing structures.

Franchising: This is different from a licensing agreement and works best for service-based businesses such as Burger King, Pizza Hut, Marriott International, Baskin-Robbins, Taco Bell, and such. The franchisee must duplicate the original business by setting up and operating under the auspices of the franchisor.

5 The Part of Tens

IN THIS PART . . .

Know the value of a whitepaper and learn how to measure ROI.

Seek out an appropriate journal, understand the masthead, learn the lingo, write a query letter, and cross your fingers and wait.

Understand the frustrations of tech writers and how to overcome them.

IN THIS CHAPTER

» Writing a compelling whitepaper

» Making your paper memorable

» Promoting and measuring ROI

Chapter 19

Ten Tips for Writing a Whitepaper

A lot of progressives really believe that if we can turn out one more whitepaper with bullet points about how to fix Problem X, we can fix it. But that's not primarily the way you reach people or move them. You reach the heart first.

—ROBERT GREENWALD, BRAVE NEW FILMS, A NONPROFIT FILM AND ADVOCACY ORGANIZATION

A *whitepaper* is an informational, persuasive, authoritative, in-depth report on a specific topic. It's intended to help learners understand an issue, solve a problem, or make a decision. Politicians use whitepapers to influence peers or constituents to think or vote a certain way. The government issues whitepapers to present policy preferences before it introduces legislation. Companies issue whitepapers to market their businesses and stand out as thought leaders in their fields. In addition, many nonprofits use whitepapers to showcase their understanding of key issues, highlight their expertise in solving complex social issues, and share their success stories.

Consider Your Audience

TECHNICAL WRITING BRIEF

Use the Technical Writing Brief in Chapter 3 and on the Cheat Sheet to understand your audience and why they should be interested in the topic. Perhaps you're presenting a common challenge they face or you're showcasing or a case study of successful industry-related companies. Don't sound salesy and don't position the paper so it sounds like a biased pitch.

Biased pitch: Ten Ways [Company] Can Save You Money on Social Media Campaigns

Unbiased whitepaper: Social Media Campaigns: Paring Marketing Needs and Platforms

TIP

The term *whitepaper* originated with the British government in 1922 when Winston Churchill issued the "British Policy in Palestine." It became known as The Churchill White Paper and responded to the riots in Jaffa. The paper confirmed that Palestine would not become a Jewish State and that Arabs would not be subordinate to Jews. (Why the term was known as "whitepaper" remains a mystery.)

Find Credible Sources

Keep a record of all your sources and give them credit. Sources can be industry experts, journals, blogs, focus groups, interviews, other whitepapers, and more. Fact-check everything. Cite your sources using annotations with footnotes or include them in a final section titled "Sources."

Include Facts and Figures

A typical whitepaper is between 2,500 and 5,000 words. It may include marketing statistics, comparison of different campaigns, complex analyses of industry trends, risk-benefit analyses, in-depth explanation of a specific process, or anything you want to subtlety promote. People respond to whitepapers more than to blatant ads (that's why I said *subtlety promote*). Whitepapers are more believable because they appear to be unbiased.

Follow a Simple Format

Create an outline (this is a must!) and liberate your inner storyteller. An outline may follow this format:

- » Title page
- » Table of contents
- » Introduction or abstract
- » Sections with headings and subheads
- » Footnotes and sources
- » Conclusion and/or call to action

Create an Eye-Catching Cover

Your cover can make the difference between a great whitepaper, a good whitepaper, or one that doesn't get read. It's the first thing readers see. It must be striking with the title taking center stage. Include a clever tagline for garnish. Choose colors that are appropriate to your industry. Example 19-1 shows four sample covers. (If you viewed them in color, you'd notice vivid shades of reds, yellows, and greens.)

EXAMPLE 19-1: Prepare an attention-grabbing cover.

kraphix/Adobe Stock

Pique Interest

CROSS REFERENCE

Every step of the way, help your learners make an informed decision. Tell learners how they'll benefit from reading your paper. Provide real-world examples. Incorporate business benefits. Pinpoint their ROI. For easy reading, break up the content with lots of headings, subheads, bullets and numbers, and graphics. Chapter 5 shows a host of examples.

Proofread and Edit Carefully

CROSS REFERENCE

Because this paper has the potential of being high-profile, it's wise to have others take a peek. Also, ensure that the readability level suits your readership. Check out Chapter 7 to learn more about readability.

Conclude with a Call to Action

Conclude with a comprehensive summary of your key points. Re-emphasize the benefits of the solution to subtly move the learners toward your company's solution. Tell learners what you want them to do and how to do it — all leading to a call to action. Don't just suggest that they leave their names and phone numbers. Offer a free trial, free assessment, gift certificate, or whatever is appropriate for your business.

Maximize Mileage

Your whitepaper can be a powerful lead-generation tool only if people know about it. Create a well-optimized landing page, use social media sites, send emails, post it on your website, include it in a blog, use pay-per-click (PPC), distribute it at workshops or events, or hire a marketing or public relations agency.

Measure the Impact

You won't know your ROI until you measure it. There are many content-marketing platforms that will host your whitepapers on your website in addition to managing lead-generation efforts and lead-scoring rules. Here are four of the more popular ones:

- Marketo (public)
- HubSpot (private)
- Eloqua (Oracle)
- Pardot (Salesforce)

IN THIS CHAPTER

» Finding the right journal

» Showing your determination

» Decoding the masthead

» Writing the query letter

» Understanding the lingo

» Avoiding the last writes

Chapter 20

Ten Tips for Publishing in a Technical Journal

If you have a dream, don't just sit there. Gather courage to believe that you can succeed and leave no stone unturned to make it a reality.

—DR. ROOPLEEN, MOTIVATIONAL WRITER

At some point in your career you may have valuable information to share. The way to reach the masses is by publishing an article in a professional publication. Whether the publication appears in print, online, or in both, the potential readership is enormous.

People who publish are part of an elite group. And the more prestigious the publication, the more the value! Not all publications pay for articles or offer honoraria; the payoff is in getting published. When you publish an article, you *add prestige to your reputation and your company's (or institution's) reputation.* You can order reprints to include in sales packets and proposals. And the best part is that you're not marketing yourself or your product; the publication is doing it for you!

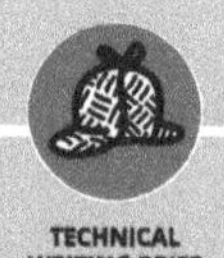

GETTING "WRITE" TO IT

Before you start to write your article, use the Technical Writing Brief in Chapter 3 and on the Cheat Sheet to understand your audience and key issues. Write your article by using the guidelines in Part 2 of this book.

If your topic is very revolutionary or controversial, you may face some obstacles in getting it published. Editors aren't necessarily crusaders. Crusaders in the publishing industry either receive journalism awards or wind up with their golden futures behind them.

A reprint of an article you write is a great promotional piece to attach to your resume or include as a handout when you present at a conference. Once you've had something appear in print, you're considered an expert.

Don't Procrastinate; Just Do It!

If you're intimidated thinking about the prospect of trying to get something published, don't be! Editors are clamoring for good material. If you ever had responsibility for publishing a newsletter, you know how difficult it is to get contributors to submit interesting articles. Editors have the same problem. The famous photos of President John F. Kennedy's assassination were submitted by an amateur photographer — one who never had anything published and happened to be in the right place at the right time. The photos and the photographer received worldwide acclaim. Submit your work. You have much to gain and nothing to lose.

If you're a newbie and have never been published, don't let that stop you. Notables such as Mark Twain, John Grisham, Mary Higgins Clark, Dr. Seuss, James Joyce, and Sheryl Lindsell-Roberts started somewhere! (I'm certainly not in their league, but I can dream.)

Don't underestimate the value of determination. Before I got my first article published, I must have sent it to every publisher in the universe. I got so many rejection letters, I could have wallpapered the Taj Mahal with them. However, I never gave up and am now a professional writer with a slew of articles and 25 books to my credit. I did it. You can too!

Hook Up with the Right Publication

You're a professional. You know what publications you and your colleagues read. If you're unsure where to submit your topic, contact professional organizations and ask for suggestions. You can also find submission information on the websites of many journals.

TIP

Before you contact the publisher, read at least a half dozen back issues of the journal you target. Doing so will give you a good idea of the type of articles the journal looks for, the style and tone, the length of the articles, and lots of other useful information.

Also visit the journal's website and check the guidelines. Many have strict guidelines about simultaneous submissions and about content, length, figures and tables, and more. Also, you may be asked to verify that your article hasn't been published or submitted elsewhere. Resubmitting an article already published in a printed or electronic journal is typically unacceptable.

Decipher the Masthead

The masthead is the front part of the publication that lists publishers, editors, phone numbers of branch offices, board of reviewing editors, member societies, and other good stuff. The only ones who read mastheads are the parents of the people mentioned and wannabe writers (who read them insatiably). At first glance, you notice that nearly everyone listed is an editor of some sort. Trying to decipher who does what is the real challenge. The best ways to find editors who accept unsolicited manuscripts are to call the publication or look online.

Understand the Lingo

Following are a few phrases you may see and what they actually mean:

- **"We don't accept unsolicited manuscripts."** Gee whiz — doesn't this sound like the publication wants to discourage freelance submissions? Not necessarily! Editors depend on submissions to keep the publication afloat. They simply want to discourage the amateur who hasn't taken the time to learn the ropes. Remember, you have nothing to lose.

- **"Reports promptly."** Yes, they may report the rejections promptly, if you hear from them at all. Editors pass around queries for review, so an acceptance may take time.
- **"Reports in two to four weeks."** Two to four months may be more like it, if at all.
- **"Pays five to ten cents a word."** You can bet you'll get five.

Write a Query Letter

A query letter is similar to a cover letter that you send with a resume. It's the first thing the editor reads. The query letter gives a clear indication of your writing style and thought process. The difference between a cover letter and query letter is that the former accompanies your resume; the latter stands alone. It's not wise to send your article unless the publication requests it. Therefore, the query letter either piques or squelches the editor's interest.

Limit your query to one page. If the title is tentative, refer to it as a "working title." Here are some things to focus on:

- Start with a hook — an attention grabber. For example, does the article present anything new or insightful?
- Stress how you intend to approach and develop the topic.
- Explain what photographs or other graphics you have to support your data. (Be mindful of copyright infringements.)
- Give your best estimate of how long the article will be (in approximate number of words, not pages).
- Specify why you're qualified to write the article. You may attach a resume or other data to support your qualifications.
- End with a request to write the article.
- Specifically ask the editor to respond.

TIP

Most submissions are online; however, some publications still use postal mail. For the latter, include a self-addressed, stamped envelope (SASE).

Follow Up after Submitting Your Manuscript

If you don't get a response within four weeks, it's appropriate to follow up with the editor again. Editors are buried under deluges of queries and may not have read yours. (You may not connect with an editor, but it may bring attention to your submission.) Once you pique the interest of the editor, your request may go before a review board. You may not be notified this is happening, but you will be notified if they accept.

Try Simultaneous Submissions

When a publication requests a copy of your manuscript with a view toward publishing it, the publication likes to feel that it has it exclusively. If more than one publisher wants to read your manuscript, put your priorities in order and send the manuscript to one publication at a time.

Don't Stress about Confidentiality

Writers are often concerned about the confidentiality of a manuscript. When you deal with a reputable publication (and you know who the big names are), you *can* trust the editor to hold your query in strict confidence. Editors won't share your writing with competing publications or disclose your ideas for building a better mousetrap. Therefore, give the editor whatever information adds strength to your query.

WARNING

Be very careful, however, about including patentable ideas, trade secrets, financial information, or anything else of a confidential nature. Before you do that, check with an attorney or an officer of your company. You may need to get written permission if you're including someone else's images, quotes, data, and so on.

Don't Take No for an Answer

The greatest number of writers — both newbies and experienced writers — have their articles rejected many, many times. Don't take it personally and have a brief pity party. Rejection may merely mean the topic isn't appropriate for that particular publication or the publication doesn't have available space at the time of your submission.

Some publications readily accept unsolicited manuscripts, some are by invitation only, and others tell you they rely on staff writers. Every publication worth its salt knows that input from a broad range of contributors strengthens its publication. Even if the publication doesn't accept unsolicited material, don't let that dissuade you from sending a query. You'll be hard to ignore if your topic is impressive and presented exquisitely.

An option is to make a few revisions and send it to another publication. Studies suggest that at least 20 percent of published articles were first rejected by another journal. If you continue to get rejections, perhaps you should put the article on hold for another time or move on.

Take the Next Steps: When Your Article Is Accepted

Congratulations! Your manuscript was accepted for publication! What can you expect next? Most correspondence between you and the journal will be done through email. The following outlines a typical process:

1. Run your victory lap!
2. Your manuscript will go to an editorial team.
3. Someone from the team will contact you with a timeline of your proof.
4. You'll have an opportunity to "yea" or "nay" any editorial changes.
5. If there are no changes, your manuscript will be ready for publication. If there are changes, there may be several back-and-forth rounds.

Once your article is published, you'll typically receive a link where you can download the PDF.

IN THIS CHAPTER

» Work overload and last-minute changes

» Issues with SMEs and micromanagers

» Hardware and software challenges

» No job security and burnout

Chapter 21

Ten Frustrations of Technical Writers

If you feel like you're losing everything [frustrated or stressed] remember that trees lose their leaves each year and they stand tall and wait for better days to come.

—KAREN SALMANSOHN, AUTHOR AND ENTREPRENEUR

According to *CNN Money*, of the tens of thousands of jobs out there, technical writing is ranked as the 13th best job. And it's the fifth least stressful job. However (as with anything in life), there is the yin and the yang — pros and woes. Frustrations rear their ugly heads even in the best jobs, and technical writing is no exception — especially when you're a newbie and haven't developed confidence.

TECHNICAL WRITING BRIEF

Before you start any writing project, fill out the Technical Writing Brief found on the Cheat Sheet. (Details are outlined in Chapter 3.) It's best to fill it out with the entire team, when possible.

Work Overload and Time Pressures

Documents need to be completed in time for product deliveries and within budgeted allowances. Some companies don't see technical writing as a priority (at least early on) and don't start the documents in time for them to be done well.

Last-Minute Changes

This is often unavoidable. There may be last-minute changes as the product continues to be developed and tested. Then proofreaders and editors are bombarding you with changes each time they're asked to peek. (Everyone in a position of giving input will do so.)

Issues with Subject Matter Experts (SMEs)

These problems can vary, depending on the communication skills of the SME.

- Poor communication skills in transferring knowledge.
- Giving vague or incomplete information and expecting technical writers to fill in large gaps.
- Not investing the time to explain concepts.
- Forgetting to keep the writers abreast of changes.
- A high-and-mighty attitude toward writers (especially toward newbies).
- Careless document review processes.

Problems with Micromanagers

Micromanagers often limit the independence of writers. New writers may benefit from guidance and feedback, but there's a big difference between guidance, feedback, and "control." For experienced writers, managers may want things done their way, even if their way may not be the right way or the best way. Contrast that with managers who offer no support — another source of frustration.

Challenges with New Products

SMEs are rarely skilled at writing technical documents for non-technical readers. Many do, however, and the results are usually less than desirable. (Ask anyone who's had trouble using documentation.) So along comes the technical writer, who can't be expert in everything. Writers are expected to grasp information instantly and write about technical products in a short period of time. This is often compounded by the fact that they often have limited access to the product(s) they're writing about. When access is limited, the writer must rely on the SME, compounding a sometimes already stressful situation.

Hardware and Software Challenges

Learning new technology can always be a challenge. This can be especially true of technical writing freelancers or vendors who walk into new companies and situations. Issues can include:

- Converting documents into other formats.
- Installing new applications and managing upgrades.
- Encountering problems with no one to troubleshoot.
- Dealing with network, file management, and other related issues.

Poorly Defined and Managed Projects

Technical writers sometimes encounter poorly defined projects, where ownership of responsibilities isn't clear and where features and deadlines are constantly in a state of flux. Projects such as these seem to hang in limbo and demand constant energy and time.

Poor Workspace Environments

Technical writers aren't always offered the most ideal workspace. Many have poorly lit areas, inadequate desks, cramped (shared) quarters, and uncomfortable chairs. This can lead to eye strain, back problems, and carpal tunnel syndrome, among other problems. Speak up *before* these issues occur.

Little or No Job Security

Job security is no guarantee in any profession. New or junior writers may be the first to be let go when a company downsizes. For freelancers/contractors, companies sometimes cancel contracts with little or no notice.

Burnout

Burnout, and these other drawbacks, don't mean you need to change your profession or flee to a deserted island. Check out Chapter 1 for options. Perhaps you need to:

- Seek out a new industry or company.
- Change the kinds of technology you write about.
- Freelance instead of working for a company (or vice versa).
- Go to the office instead of working at home (or vice versa).
- Offer an adult education course in technical writing.
- Get into streaming, AI, UX, or some other specialized niche.
- Write a book — but don't compete with this one (wink, wink).

CROSS REFERENCE

Take a gander at the end of Chapter 1 to learn how I turned technical writing into a lucrative, multi-faceted career. There's no reason you can't do the same!

TIP

KEEPING FRUSTRATIONS AT BAY

When you're feeling particularly frustrated, keep these simple gems in mind to cool your head and steady your heart:

- Don't commit to more than you can handle.
- Plan, prioritize, and keep yourself organized.
- Don't procrastinate.
- Ask for help when needed — and know whom to ask.
- Speak to a trusted colleague or manager if frustrations get out of hand (due to no fault of your own).
- Take a break! Get some coffee! Go outside and take a quick, brisk walk! Check in with beloved friends, family, and pets! There's more to life than your job.

Appendix A
Punctuation Made Easy

Punctuation is one of the most significant tools you have to create documents that bear the mark of your own voice. When you speak aloud, you constantly punctuate sentences with your voice and body language. And when you write, you make a sound in the reader's head. Your "writing voice" can be a dull, sleep-inducing mumble (like a tedious, unformatted document) or it can be a joyful sound, a shy whisper, a throb of passion. It all depends on the punctuation you use.

SHERYL SAYS

I present the punctuation marks in the order in which they're most commonly used and confused.

GENERAL GUIDELINES

Here are a few general punctuation guidelines:

- **Place commas and periods inside quotation marks.**

  The engineer wanted to hear her supervisor say "yes."

- **Place semicolons and colons outside quotation marks.**

  The design used is "Taguchi L16c"; it consists of 16 wafers.

(continued)

(continued)

- **Place question marks and exclamation points inside the quotes only when they apply to the quoted material.**

  "Did you read Sally's findings?" Jim asked.

- **Place question marks and exclamation points outside the quotes when they apply to the entire sentence.**

  Did the supervisor say, "Set the pressure to 10 psi"?

You can also use punctuation to stress what you want your readers to see as important. For example, the following three sentences are worded identically. Yet the different marks of punctuation give each a unique sound:

Dashes: The Ace Chemical Company — winner of the annual Service Award — just introduced its new product line. (The dashes emphasize the award.)

Parentheses: The Ace Chemical Company (winner of the annual Service Award) just introduced its new product line. (The parentheses downplay the award and emphasize the introduction of a new product line.)

Commas: The Ace Chemical Company, winner of the annual Service Award, just introduced its new product line. (The commas neutralize the entire sentence.)

SHERYL SAYS

One rule of thumb is, remember to be consistent. For example, if you decide to put a comma before the final item in a series, do that regularly.

Commas

Commas are the most frequently used (and misused) punctuation mark. While periods indicate a *stop* in thought, commas act as *slow signs* — like speed bumps. They let you know which items are grouped together, what's critical to the meaning of the sentence, and more — as the general rules listed here indicate.

- **Use commas to separate three or more items in a series.** A comma after the final item is optional. It can, however, increase clarity. The choice is yours, but be consistent.

  We'll need cartridges, staples, and copy paper.

- **Use a comma before a conjunction *(and, but, or, nor, for, so,* or *yet)* that joins what could be two complete sentences.**

  Joining two sentences: Joe recognized the four presenters, but he couldn't recall their names.

  One sentence plus a clause: Joe recognized the four presenters but couldn't recall their names.

- **Don't place a comma before *because.***
- **Use commas to separate items in an address or a date.** But don't use any punctuation before a ZIP code.

  On Monday, April 2, XXXX, Cranston Technology Co. will move to 13 James Street, Maxinkuckee, IN 46511.

- **Use commas to set off an expression that explains or modifies the preceding word, name, or phrase.**

  Jim Smith, our IT specialist, will be on vacation the week of May 3.

- **Use commas to set off one or more words that directly address the person to whom you're speaking by name, title, or relationship.**

  Please let me know, Marv, if you can add anything to those findings.

- **Use a comma after an introductory phrase if it's followed by a complete sentence. This type of clause may include introductory words such as *when, if, as,* and *although.***

  If we state our case clearly, we should get the funding.

- **Use commas around a phrase that isn't necessary to the meaning of the sentence.**

  Barbara, whom you met at the office last week, is a speaker at the semiconductor conference. (Barbara is a speaker at the conference regardless of when you met her.)

- **Don't place commas around information that makes the sentence clear.**

  The person who meets all our qualifications will never be found.

- **Use commas to set off expressions that interrupt the natural flow of the sentence.** These expressions include *as a result, in fact, therefore, however, consequently, for example, in fact, on the contrary,* and others.

  We will, therefore, continue with the project.

- **Use commas to clarify a sentence that would otherwise be confusing.**

  It may be a long, long time before we get the test results.

In 1999, 53 computers were sold. (Or: In 1999, fifty-three computers were sold.)

Without Bill, Jim can't proceed.

Only three weeks before, I had lunch with him.

» **Use commas for emphasis.**

The shipment, unfortunately, was delayed.

» **Use commas to show contrast.**

The assignment is long, but not difficult.

» **Use commas to identify a person who is quoted directly.**

John Naisbitt, American business writer and social researcher, said, "We are drowning in information but starved for knowledge."

» **Use commas to set off designations, titles, and degrees that follow a name.**

Ted Adler, President of Verdox Company, will be next month's speaker.

» **Use a comma to divide a sentence that starts as a statement and ends as a question.**

I can't think of anything further, can you?

» **Use commas to separate items in reference material.**

You can find the Peano-Gosper Curve in *The Fractal Geometry of Nature,* by Benoit B. Mandelbrot, Chapter 7, page 70.

» **Use a comma to separate words when the word *and* is omitted.**

Please include a stamped, self-addressed envelope.

ONE SPACE, NOT TWO

In the olden days of typewriters, spacing twice after a punctuation mark was sound advice. After all, doing so was the only way to clearly separate one sentence from another. With computers, however, we have proportional spacing that shows a clear distinction between sentences.

Therefore, you should *space once* after a period, colon, exclamation point, question mark, quotation mark, or any other mark of punctuation that ends a sentence.

Colons and Semicolons

Semicolons are separators that are stronger than commas and weaker than periods. Colons direct the reader's attention to what follows. The following sections show how to use them correctly.

Semicolons

Consider semicolons a cross between periods and commas. They create more pause than commas, yet less than periods. Following are some rules about when to use them:

- **Use a semicolon in place of a conjunction *(and, but, or, nor, for, so,* or *yet)* to join complete sentences.**

  The Georgia plant supplies the raw material; the Chicago plant provides the finished product.

- **Use a semicolon when a parenthetical word *(however, therefore,* and the like) or phrase introduces a separate sentence.**

  The project came to a standstill during the strike; however, we did eke out a small profit.

- **Use a semicolon to separate items in a series when the items themselves have commas.**

  The milestones were January 15, 2000; February 20, 2000; and April 15, 2000.

Colons

Colons are marks of anticipation. They serve as introductions and alert you to a close connection between what comes before and after it.

- **Use a colon after an introduction that includes or implies *the following* or *as follows.***

  These are the people you will meet: James Smith, Jerry Alexander, and Bob Nethers.

- **Use a colon to introduce a long quotation.**

  Professor Longwinded said: "The project came to a standstill after . . ."

- **Use a colon to separate hours and minutes.**

  He should arrive at 10:45 a.m.

Dashes and Parentheses

Dashes and parentheses affect how readers understand information. Dashes highlight the text; parentheses play down the text.

Dashes

Dashes (often considered strong parentheses) are vigorous and versatile. They can stand alone or be used in pairs. Just don't overdo dashes, or they lose their impact.

TIP

Here are two ways to form an em dash: If you don't space between the second hyphen and the word that follows, the two hyphens magically become an em dash. Or in Word you can choose the Insert menu and then highlight Symbols. Some writers leave one space before and after the em dash; others leave no spaces. Check your company's style guide. If there isn't one, pick the one you think looks best.

- **Use dashes to set off expressions you want to emphasize.**

  This application — as unbiased tests have disclosed — is more powerful than what you're currently using.

- **Use a dash to indicate a strong afterthought that disrupts the sentence.**

  I know you're looking for — and I hope this helps — a list of qualified people.

Parentheses

Parentheses (often considered weak dashes) are like a sideshow; they're used to enclose one or more words in a sentence that aren't essential to the meaning of the sentence. Some examples of when to use parentheses follow:

- **Use parentheses around an expression that you want to de-emphasize.**

  A parenthetical expression is one that doesn't change the meaning of the sentence — that is, removing the expression doesn't alter the gist.

  This application (as unbiased researchers have established) is more powerful than what you're using.

- **Use parentheses around references to charts, pages, diagrams, authors, and so on.**

  Please read the section on fossils (pages 36-52).

» **Use parentheses to enclose numerals or letters that precede items in a series.**

We are hoping to (a) get the draft completed by May 5, (b) get feedback by May 8, and (c) go to press on June 5.

REMEMBER

When you enclose a sentence in parentheses, add the period before the closing parenthesis. Likewise, if the information in the parentheses is not a full sentence, the punctuation goes outside of the ending parenthesis (like this).

Brackets

Brackets aren't substitutes for parentheses. They have their own place in the world, as the following guidelines explain:

» **Use brackets to enclose words that you add to a direct quote.**

He said, "The length of the study [from January to November] was entirely too long."

» **Use brackets as parentheses within parentheses.**

Your order (including one dozen blue pens [which aren't available], three dozen green pens, and five dozen red pens) will ship on Monday, September 8.

Other Punctuation

Still to come are quotation marks, apostrophes, ellipses, hyphens, question marks, exclamation marks, periods, and slashes.

Quotation marks

Quotation marks are reserved for those occasions when you're citing something verbatim. If you paraphrase, don't use quotation marks.

Quoting: Mr. Ramirez said, "Please come to the meeting at 2:00."

Paraphrasing: Mr. Ramirez asked her to come to the 2:00 meeting.

» **Use quotation marks to enclose direct quotes.**

"The high tech industry is vital to the economy," said the CEO.

» **Use quotation marks to enclose articles from magazines, songs, essays, short stories, one-act plays, sermons, paintings, lectures, and so on.**

A recent issue of *Physics Today* magazine contained an article "Career Opportunities in Optics."

» **Use quotation marks to set off words or phrases introduced by expressions (such as the word, known as, was called, marked, entitled, and so on).** Another option is to use italics.

Quotes: See the drawing marked "Wafer configure utilization."

Italics: See the drawing marked *Wafer configure utilization.*

TIP

» **Use single quotation marks around a quotation within a quotation.**

The consultant said, "You would do well to heed Mr. Smith's advice: 'Give the public what it wants, and you will be in business for a long time.'"

Ellipses

Ellipses show that words or names are omitted in a quotation. Place the ellipses where the omission occurs. They're formed by typing three periods with a space between each set of periods. When ellipses end a sentence, you don't need a final period.

Omission at the beginning: ". . . The experiment consisted of printing 24 wafers at pre-determined screenprint settings."

Omission at the end: "A Fractional Factorial 2^3 blocked by room temperature was used. The experiment consisted of printing 24 wafers at predetermined screenprint settings . . ."

Omission somewhere in between: "The experiment consisted of printing 24 wafers . . . for shorts, opens, and solderbump height."

Apostrophes

Apostrophes aren't flying commas; they show possession or omissions.

Possession

Possession refers to ownership, authorship, brand, kind, or origin. These guidelines demonstrate how to use apostrophes to show possession:

» **Apostrophes are used most commonly with nouns to show possession.**

In the following sentences, which host would you prefer?

The presenter called the guests names when they arrived.

The presenter called the guests' names when they arrived.

» **Form the possessive case of a singular noun by adding an apostrophe.**

The idea is Jim's brainchild.

» **Form the possessive of a regular plural noun (one ending in *s*) by adding an apostrophe after the *s*.**

The Murphys' lab is closed.

» **Form the possessive of an irregular plural noun (one not ending in *s*) by adding an apostrophe and *s*.**

The salespeople's territories are being divided.

» **To show joint ownership, add the apostrophe and *s* after the last noun. To show single ownership, add the apostrophe and *s* to each noun.**

Joint ownership: Jim and Pat's company is issuing an IPO.

Individual ownership: Jim's and Pat's lockers are on separate floors.

» **In hyphenated words, put the apostrophe at the end of the possession.**

He borrowed his brother-in-law's computer.

» **To make an abbreviation possessive, put an apostrophe and *s* after the period. If the abbreviation is plural, place an apostrophe after the *s*.**

The Smith Co.'s testing starts next month.

Two M.D.s' opinions are needed.

» **Express time and measurement in the possessive case.**

We'll have an answer in one week's time.

TIP

It's becoming commonplace to write the names of companies and publications without apostrophes. When in doubt, check it out.

Omission

Use apostrophes to show that letters (as in contractions) or numbers (as in *the '90s*) are missing. Some companies frown on using contractions and prefer the more formal style. Defer to the style of the company.

» **Use an apostrophe to form the plural of a number, letter, or symbol, or word used as a separate word.**

His last name has two f's.

Sally doesn't always pronounce her r's at the end of a word.

No if's, and's, or but's.

» **Use an apostrophe to show possession of initialisms or acronyms.** Some writers eliminate the apostrophe when there's little chance of misreading.

I used MPR's suggestions.

Hyphens

Don't confuse hyphens (-) with em dashes (—). They're different species. Hyphens function primarily as spelling devices.

» **Use a hyphen to join compound words that come before a noun.** Compound words are two or more words used as a unit to describe a noun.

The company gave the researchers a ten-day extension.

The company gave the researchers an extension of ten days.

» **Use a hyphen for compound numbers and written-out fractions.**

One hundred fifty-two people attended the meeting.

This is three-fourths the annual revenue.

» **Use a hyphen between a prefix that ends with a vowel and a word beginning with the same vowel. (When in doubt, check it out.)**

It was said to be a pre-existing condition.

The television station pre-empted my favorite program.

Question marks

Question marks serve as stop signs. Although you probably use them correctly, there are a few tricky situations. Hopefully, these guidelines can help demystify them:

» **Use a question mark after a short, direct question that follows a statement.**

You saw the requisition, didn't you?

» **Use a question mark after each item in a series of questions within the same sentence.**

Which of the candidates has the most experience? Mary? Joe? Jeff?

Another option is the following: Mary, Joe, or Jeff?

Don't capitalize the questions unless the beginning word should be capitalized, such as in the first of the above examples where there are names.

» **Use a question mark enclosed in parentheses to express doubt.**

He said the results are due on April 8. (?)

Exclamation points

Exclamation points are reserved for words or thoughts that show strong feelings or emotions, as the following examples demonstrate:

Please try to do better!

That was an inspiring talk. Congratulations!

Periods

Periods are the stop signs of punctuation. They slow you down before you go on to the next thought.

» **Use a period after a statement, command, or request.**

Thank you for getting us the results so promptly.

» **Use a period after words or phrases that logically substitute for a complete sentence.**

No, not at all.

» **Use periods when writing abbreviations, acronyms, or initialisms.**

A number of dictionaries, however, are citing many — for example, IBM, FDIC, and CPA — without periods. When in doubt, check it out. For more information about abbreviations, acronyms, or initialisms, see the sidebar in Appendix C.

Slashes

These critters go by a variety of names: slant lines, virgules, bars, or shilling lines. They separate or show an omission, such as in care of (c/o) or without (w/o).

- **Use slashes in *and/or* expressions.**

  The IS/training departments will present the training.

- **Use slashes in Internet addresses.**

  `https://www.dummies.com/`

Appendix **B**

Grammar's Not Grueling

You may remember, when you were a kid, asking your mother, "Mom, can Pat and me go to the movies?" Your mother replied, "That's *may Pat and I*" — and she didn't give you the money until you corrected your grammar. Although you didn't think so then, your mother was doing you a favor. Poor grammar didn't get you far with your mother, and it doesn't get you far in the business world.

Testing Your Grammatical Skills

Take a look at the following sentences and see if you notice any errors. If you find all the mistakes, you're a grammar guru. You find the answers at the end of the chapter.

1. A group of 75 computer specialists are waiting for the test results.
2. With most of the votes counted, the winner was thought to be her.
3. What was the name of the speaker we had yesterday?
4. Dr. Allen, who specializes in kinetics, would certainly be interested if he was here now.
5. The MVC Technology Company is celebrating their 50th anniversary.

Adjectives

Adjectives answer at least one of the following questions: *What kind? Which? What color? How many?* or *What size?* They're words, phrases, or clauses that modify, describe, or limit the noun or pronoun they describe. You can use adjectives to transform an ordinary sentence into a tantalizing one. This can be a nice touch in a technical document that may be otherwise dry.

Forms of adjectives

Adjectives take different forms, depending on the noun or nouns they modify.

- **Use a positive adjective when you're not comparing anything.**

  It's *warm* in the lab.

- **Use a comparative adjective when you're comparing two things.**

  It's *warmer* in the lab today than it was yesterday.

- **Use a superlative adjective when you're comparing three things or more.**

  It's the *warmest* the lab has been all week.

Absolute adjectives

Some adjectives are absolute; they either *are* or *aren't*. For example, one thing can't be rounder than something else. Either it's round or it's not. Following are some adjectives that are considered absolute:

Complete	Correct	Dead	Empty
Genuine	Parallel	Perfect	Right
Round	Stationary	Unanimous	Wrong

TIP

Express the comparative and superlative forms of absolute adjectives by adding "more nearly" or "most nearly." For example, Jason's assumption was more nearly correct than Jim's.

Compound adjectives

In many cases, you use a hyphen to join together two adjectives so they form a single description. Use a hyphen only when the compound adjective comes before the noun, not after.

> ***Before the noun:*** A part-time job; a two- or three-year experiment
>
> ***After the noun:*** A job that's part time; an experiment of two or three years

Here are two exceptions:

1. Eliminate the hyphen when you generally think of the words as a unit; for example, post office address, life insurance, word processing, and the like.
2. Don't put a hyphen between adjectives if the first one ends in *-est* or *-ly*; for example, newly elected officer, freshest cut flowers, and so forth.

Articles

Use *the* to refer to a specific item and *a* or *an* to a non-specific item. Use *a* when a consonant sound follows the *a* (*a* vector, *a* method); use *an* when a vowel sound follows (*an* equation, *an* inductor).

Adverbs

Just as adjectives add pizzazz to nouns, adverbs spice up verbs. Adverbs modify verbs, adjectives, or other adverbs. They answer one or more of these questions: *How? When? Why? How much? Where?* or *To what degree?* Adverbs take different forms for the positive, comparative, and superlative, just as adjectives do.

Adjectives ending in *-ly* may also function as adverbs — depending on what they're modifying.

> ***Adjective:*** The professor's handwriting is *legible.*
>
> ***Adverb:*** The professor writes *legibly.*

Conjunctions

Conjunctions connect two or more words, phrases, or clauses that are equal in construction and importance. Common conjunctions are *and, or, for, so, but, nor,* and *yet.*

CROSS REFERENCE

For information about punctuating sentences that contain conjunctions, see Appendix A.

Correlative conjunctions

You use some conjunctions in pairs to join the elements of a sentence. The most often-used pairs are the following:

Both/and	So/as	Not only/but also	Whether/or
Either/or	Neither/nor	As/as	Whether/or not

Subordinate conjunctions

Clauses starting with the following words and phrases always function as adverbs, adjectives, or nouns.

After	Although	As	As if	As long as	As though
Because	Before	Even if	Except	If	In order that
Provided	Since	Than	That	Though	Unless
Until	When	Where	While		

Double Negatives

If you ever said, "I don't want no liver," what you said is that you do want liver. Two negatives equal a positive. Never use two negative words to express one positive idea.

Correct: I don't have *any* solutions.

Incorrect: I don't have *no* solutions.

Nouns

Nouns — although critical to every sentence — are probably the least sexy part of speech. They don't create any emotion or add flair to your thoughts; they're merely *people, places,* or *things*. Proper nouns are specific and capitalized. Common nouns aren't specific or capitalized.

Proper nouns: New York City; New York University; Main Street

Common nouns: The city, the university, the street

Collective nouns are groups. When groups act as units (companies, councils, audiences, faculties, unions, teams, juries, committees, and so on), use a singular verb. When members of the group act independently, use a plural verb.

Acting collectively: The team *is* going to test the equipment.

Acting individually: The team *are* conducting separate tests.

Prepositions

Prepositions show the relationship between words and sentences. Here are some common prepositions:

Above	About	Across	After	Along
Among	Around	At	Before	Behind
Below	Beneath	Beside	Between	Beyond
By	Down	During	Except	For
From	In	Inside	Into	Like
Near	Of	Off	On	Since
To	Toward	Through	Under	Until
Up	Upon	With	Within	

Pronouns

TIP

The language of genders has evolved over time. Many of us have become accustomed to using she/her/hers for females and he/his/him for males. As gender vocabulary continues to evolve, it's proper to address a singular person as they, them, ze, or hir. Inclusive language offers respect, safety, and belonging to all people.

Pronouns are words that substitute for nouns. Pronouns normally agree with the nouns they replace in person, number, and gender.

- **Use *I, you, he, she, it, we,* or *they* when the pronoun is the subject of the sentence, or when it follows any form of the verb *to be: be, am, is, are, was, were, been, being, will be, had been*, and so on.**

  Is Sam there? Yes, this is *he.*

  They prefer to be called Logan now.

- **Use *me, you; him, her, it, us,* or *them* when the pronoun is the object of either the verb or a preposition. These pronouns tell you "who" or "what."**

  I will call *her* when the magazine arrives.

  I told *them* to have a seat.

REMEMBER

  One trick to use when two pronouns come together is to break the thought into two sentences. It works every time!

  Correct: *Tom and I* participated in the experiment.

  (Tom participated in the experiment. I participated in the experiment.)

  Incorrect: *Tom and me* participated in the experiment.

- **Use a possessive pronoun to indicate possession, kind, origin, brand, authorship, and so on. The possessive pronouns are *my, mine, your, yours, his, her, hers, its, our, ours, their,* and *theirs.***

  The decision is completely *his.*

  The coffee lost *its* taste. ("It's" is only used when you mean "it is.")

A common error occurs when referring to nouns such as in the word *company.* Because company is a singular noun, the pronoun that replaces the company name must be singular.

Correct: The company is having *its* annual meeting on August 15.

Incorrect: The company is having *their* annual meeting on August 15.

REMEMBER

As mentioned in Chapter 6, many companies have opted to no longer alternate between the use of "he" and "she" to avoid favoring one gender over the other. Instead, it's better to use the singular "they" when referring to individuals. Other ways to avoid gendered pronouns include using plural nouns, repeating the noun, and using the pronouns "one" or "who," when appropriate. When a person's gender and pronoun preference is known or established (such as for historical and public figures), it's appropriate to use the relevant pronoun.

Singular pronouns

Certain pronouns are always singular and take singular verbs and pronouns, including *anybody, anyone, anything, each, either, everybody, everyone, everything, much, neither, nobody, nothing, one, somebody, someone,* and *something.*

Everyone, including Pete and Fatima, *is* going to the party.

Neither Carmine's nor Kwame's proposals *is* acceptable.

Who and whom

Some people try to mumble these words and hope that listeners won't notice their indecision. Writers don't have that luxury. But it is possible to think of *who* and *whom* in easy terms!

When you can substitute *he, she,* or *they,* use *who.* And when you can substitute *her, him,* or *them,* use *whom.* (For the latter, I generally plug in an *-m* ending.)

The company needs a person *who* knows the new software. (*He* or *she* knows the new software.)

Are you the person to *whom* I spoke yesterday? (I spoke to *her* or *him.*)

Verbs

Verbs are the most important part of sentences because they express actions, conditions, or states of being. Verbs make statements about the subjects and can breathe life into dull text.

Gerunds

Gerunds are words or phrases whose roots are verbs ending with *-ing*. Gerunds start out as verbs, but act as nouns. When gerunds are preceded by nouns or pronouns, the nouns or pronouns take the possessive form.

> I don't like *your giving* me such short notice.
>
> *Ted's yelling* is quite irritating.

Dangling participles

If your participles dangle, it's nothing to be ashamed of. The condition's curable. *Dangling participles* are nothing more than verbs that don't clearly or logically refer to the nouns or pronouns they modify. Participles can dangle at the beginning or end of sentences. The following shows how to "undangle" participles:

> ***Correct:*** While Denzel attended the meeting, the computer malfunctioned.
>
> ***Incorrect:*** While attending the meeting, the computer malfunctioned. (Who attended the meeting? This sentence implies that the computer attended the meeting because that thought dangles.)

Were and was

Have you ever fantasized about being someone else? The English language provides a verb for those fantasies. "I wish I *were* . . ." The verb *were* is often used to express wishful thinking or an idea that's contrary to fact. *Was*, on the other hand, indicates a statement of fact.

> Fiona acts as if she *were* president of the company. (Wishful thinking.)
>
> If Aki *was* at the conference, I didn't see them. (They may have been there.)

REMEMBER

Was is the past tense of *is*. Why am I mentioning the obvious? Because people often mistakenly use *was* for the present tense when referring to something that's already happened.

> ***Correct:*** I thoroughly enjoyed reading the report — even though it *is* 950 pages long.
>
> ***Incorrect:*** I thoroughly enjoyed reading the report — even though it *was* 950 pages long.

Subject and verb agreement

One of the most basic rules in grammar is that the subjects and verbs of sentences must agree. Both must be singular or both must be plural. Although most situations are pretty straightforward, the following sentences demonstrate situations that may be a little tricky:

» **Don't be fooled by interrupting phrases.**

Correct: The software, despite the new installation manuals, still *takes* several days to install. (The subject is *software*.)

Incorrect: The software, despite the new installation manuals, still *take* several days to install.

» ***A, many, an, each,* and *every* always take a singular verb.**

Each and every computer *has* a modem.

Many a person *is* denied this chance.

» ***None, some, any, all, most,* and *fractions* are either singular or plural, depending on what they modify.**

Half the shipment *was* misplaced. (The subject, *shipment,* is singular.)

Half the boxes *were* misplaced. (The subject, *boxes,* is plural.)

» **When referring to the name of a book, magazine, song, company, or article, use a singular verb even though the name may be plural.**

Little Women is a great classic.

Wanderman & Greenberg *is* a fine team of attorneys.

» **When referring to an amount, money, or distance, use a singular verb if the noun is thought of as a single unit.**

I think *$900 is* a fair price.

There *are 10 yards* of wire on the reel.

» **When *or* or *nor* is used to connect a singular and plural subject, the verb must agree in number with the person or item that is closest to the verb.**

Neither Kamau nor the *assistants were* available.

Neither the assistants nor *Kamau was* available.

Answers to the Quickie Quiz

1. A group of 75 computer specialists *is* waiting for the test results. (*Group* is a singular subject and takes a singular verb.)
2. With most of the votes counted, the winner was thought to be *she.* (*She* was thought to be the winner. The nominative case is used when the pronoun is the subject of the sentence or when it follows any form of the verb *to be.*)
3. What *is* the name of the speaker we had yesterday? (The speaker's name hasn't changed. Why use the past tense?)
4. Dr. Allen, who specializes in kinetics, would certainly be interested if he *were* here now. (Not a statement of fact. He isn't here now.)
5. The MVC Technology Company is celebrating *its* 50th anniversary. (The MVC Technology Company is a singular subject and takes a singular verb.)

Appendix C
Abbreviations and Metric Equivalents

In today's multinational and multicultural world, it's wise to identify abbreviations and include both U.S. customary units and metric units. This appendix shows many that are commonly used and how to apply them. It also has tables of chemical elements, postal abbreviations, and metrics and U.S. equivalents.

Writing Abbreviations

Write out names or expressions the first time they appear. Thereafter, use abbreviations. If you're absolutely sure the reader will understand your reference, however, there's no need to write it out. When you do need to explain your reference, here's how to present it:

> *Write out random access memory (RAM) the first time you mention it in a document. Thereafter, you may use RAM because you already identified the reference for the reader.*

If you deliver a paper document to a wide range of readers, some of whom may not understand your abbreviations, consider including a glossary of terms and abbreviations at the end of your document.

Treat the company or organization name as the company treats it. For example, if a company uses *Company*, don't write *Co.* Check the company website or letterhead for accuracy. If you can't check it out, write it out.

ACRONYMS AND INITIALISMS

What's the difference between acronyms and initialisms? Acronyms are formed by combining the first letter of several words and pronouncing the string of letters as a single word. Initialisms are also formed by combining the first letter of several words, but they're pronounced as separate letters.

Acronyms (pronounced as words):

- SCORM (Shareable Content Object Reference Model)
- LASER (Light Amplification by Stimulated Emission of Radiation)
- COTS (Commercial Off-the-Shelf Software)

Initialisms (said as individual letters):

- OCR (Optical Character Recognition)
- LMS (Learning Management System)
- ROI (Return on Investment)

In business, industry, education, and government, acronyms and initialisms are often used by people who work together. That's fine as long as the readers easily understand your frame of reference, but it's quite possible that the reference may not be known to those outside your magical kingdom. In fact, certain acronyms and initialisms mean different things to different professionals. For example, following are a few associations that "ABA" may represent:

- American Banking Association
- American Bar Association
- American Booksellers Association
- American Bowling Association
- Association for Behavioral Analysis

The bottom line? When in doubt, spell it out (at least on first reference).

Chemical elements

Table C-1 shows the chemical elements, their symbols, and their atomic numbers.

TABLE C-1 Chemical Elements

Name	Symbol	Atomic No.	Name	Symbol	Atomic No.
Actinium	**Ac**	89	Barium	**Ba**	56
Aluminum	**Al**	13	Berkelium	**Bk**	97
Americium	**Am**	95	Beryllium	**Be**	4
Antimony	**Sb**	51	Bismuth	**Bi**	83
Argon	**Ar**	18	Boron	**B**	5
Arsenic	**As**	33	Bromine	**Br**	35
Astatine	**At**	85	Cadmium	**Cd**	48
Calcium	**Ca**	20	Iron	**Fe**	26
Californium	**Cf**	98	Krypton	**Kr**	36
Carbon	**C**	6	Lanthanum	**La**	57
Cerium	**Ce**	58	Lawrencium	**Lr**	103
Cesium	**Cs**	55	Lead	**Pb**	82
Chlorine	**Cl**	17	Lithium	**Li**	3
Chromium	**Cr**	24	Lutetium	**Lu**	71
Cobalt	**Co**	27	Magnesium	**Mg**	12
Copper	**Cu**	29	Manganese	**Mn**	25
Curium	**Cm**	96	Mendelvium	**Md**	101
Dysprosium	**Dy**	66	Mercury	**Hg**	80
Einsteinium	**Es**	99	Molybdenum	**Mo**	42
Erbium	**Er**	68	Neodymium	**Nd**	60
Europium	**Eu**	63	Neon	**Ne**	10
Fermium	**Fm**	100	Neptunium	**Np**	93
Fluorine	**F**	9	Nickel	**Ni**	28
Francium	**Fr**	87	Niobium	**Nb**	41

(continued)

TABLE C-1 *(continued)*

Name	Symbol	Atomic No.	Name	Symbol	Atomic No.
Gadolinium	**Gd**	64	Nitrogen	**N**	7
Gallium	**Ga**	31	Nobelium	**No**	102
Germanium	**Ge**	32	Osmium	**Os**	76
Gold	**Au**	79	Oxygen	**O**	8
Hafnium	**Hf**	72	Palladium	**Pd**	46
Helium	**He**	2	Phosphorus	**P**	15
Holmium	**Ho**	67	Platinum	**Pt**	78
Hydrogen	**H**	1	Plutonium	**Pu**	94
Indium	**In**	49	Polonium	**Po**	84
Iodine	**I**	53	Potassium	**K**	19
Iridium	**Ir**	77	Praseodymium	**Pr**	59
Promethium	**Pm**	61	Terbium	**Tb**	65
Protactinium	**Pa**	91	Thallium	**Tl**	81
Radium	**Ra**	88	Thorium	**Th**	90
Radon	**Rn**	86	Thulium	**Tm**	69
Rhenium	**Re**	75	Tin	**Sn**	50
Rhodium	**Rh**	45	Titanium	**Ti**	22
Rubidium	**Rb**	37	Tungsten	**W**	74
Ruthenium	**Ru**	44	Unnilhexium	**Unh**	106
Samarium	**Sm**	62	Unnilpentium	**Unp**	105
Scandium	**Sc**	21	Unnilquadium	**Unq**	104
Selenium	**Se**	34	Unnilseptium	**Uns**	107
Silicon	**Si**	14	Uranium	**U**	92
Silver	**Ag**	47	Vanadium	**V**	23
Sodium	**Na**	11	Xenon	**Xe**	54
Strontium	**Sr**	38	Ytterbium	**Yb**	70

Name	Symbol	Atomic No.	Name	Symbol	Atomic No.
Sulfur	**S**	16	Yttrium	**Y**	39
Tantalum	**Ta**	73	Zinc	**Zn**	30
Technetium	**Tc**	43	Zirconium	**Zr**	40
Tellurium	**Te**	52			

Postal abbreviations

The United States Postal Service requests that you use the two-letter state abbreviation in all mailings. Don't use periods. Table C-2 shows the abbreviations for the United States and its territories.

TABLE C-2 United States and Territories Postal Abbreviations

State	Postal Code	State	Postal Code
Alabama	AL	Montana	MT
Alaska	AK	Nebraska	NE
Arizona	AZ	Nevada	NV
Arkansas	AR	New Hampshire	NH
California	CA	New Jersey	NJ
Canal Zone	CZ	New Mexico	NM
Colorado	CO	New York	NY
Connecticut	CT	North Carolina	NC
Delaware	DE	North Dakota	ND
District of Columbia	DC	Ohio	OH
Florida	FL	Oklahoma	OK
Georgia	GA	Oregon	OR
Guam	GU	Pennsylvania	PA
Hawaii	HI	Puerto Rico	PR

(continued)

TABLE C-2 *(continued)*

State	Postal Code	State	Postal Code
Idaho	ID	Rhode Island	RI
Illinois	IL	South Carolina	SC
Indiana	IN	South Dakota	SD
Iowa	IA	Tennessee	TN
Kansas	KS	Texas	TX
Kentucky	KY	Utah	UT
Louisiana	LA	Vermont	VT
Maine	ME	Virginia	VA
Maryland	MD	Virgin Islands	VI
Massachusetts	MA	Washington	WA
Michigan	MI	West Virginia	WV
Minnesota	MN	Wisconsin	WI
Mississippi	MS	Wyoming	WY
Missouri	MO		

Metric and U.S. Equivalents

TIP

If your primary audience is in the United States, place the metric conversion in parentheses directly after the U.S. customary unit:

> Jia Li used about 1 gallon (3.79 liters) of gas driving to the convention.

If your primary audience is outside of the United States, do the reverse:

> Jia Li used about 3.79 liters (1 gallon) of gas driving to the convention.

The United States is nearly the only country that doesn't use the metric system (Liberia and Myanmar are two others). C-3 through C-7 show some of the popular metric measurements and their U.S. equivalents.

TABLE C-3

Linear Measures

U.S. System	Metric System
1 inch	25.4 millimeters (mm)
1 inch	2.54 centimeters (cm)
1 foot	304.8 millimeters (mm)
1 foot	30.48 centimeters (cm)
1 foot	0.3048 meter (m)
1 yard (36 in.; 3 ft.)	0.9144 meter (m)
1 rod (16.5 ft.; 5.5 yds.)	5.029 meter (m)
1 statute mile (5,280 ft.; 1760 yds.)	1,609.3 meters (m)
1 statute mile (5,280 ft.; 1760 yds.)	1.6093 kilometers (km)
U.S. System	Metric System
1 millimeter (mm)	0.03937 in.
1 centimeter (cm)	0.3937 in.
1 meter (m)	39.37 in.
1 meter (m)	3.2808 ft.
1 meter (m)	1.0936 yds.
1 kilometer (km)	3,280.8 ft.
1 kilometer (km)	1,093.6 yds.
1 kilometer (km)	0.62137 mi.

TABLE C-4

Liquid Measures

U.S. System	Metric System
1 fluid ounce (fl. oz.)	29.673 milliliters (ml)
1 pint (16 fl. oz.)	0.473 liter (l)
1 quart (2 pints; 32 fl. oz.)	9.4635 deciliters (dl)
1 quart (2 pints; 32 fl. oz.)	0.94635 liter (l)
1 gallon (4 quarts; 128 fl. oz.)	3.7854 liters (l)

(continued)

TABLE C-4 *(continued)*

U.S. System	Metric System
1 milliliter (ml)	0.033814 fl. oz.
1 deciliter (dl)	3.3814 fl. oz.
1 liter (l)	1.0567 qt.
1 liter (l)	0.26417 gal.

TABLE C-5

Area Measures

U.S. System	Metric System
1 square inch (0.007 sq. ft.)	6.452 square centimeters (cm^2)
1 square inch (0.007 sq. ft.)	645.16 square millimeters (mm^2)
1 square foot (144 sq. in.)	929.03 square centimeters (cm^2)
1 square foot (144 sq. in.)	0.092903 square meters (m^2)
1 square yard (9 sq. ft.)	0.83613 square meters (m^2)
1 square mile (640 acres)	2.59 square kilometers (km^2)
U.S. System	Metric System
1 square millimeter (mm^2)	0.00155 square inches (sq. in.)
1 square centimeter (cm^2)	0.155 square inches (sq. in.)
centiare	10.764 square feet (sq. ft.)
square kilometer (km^2)	0.38608 square miles (sq. mi.)

TABLE C-6

Capacity

U.S. System	Metric System
1 cubic inch (0.00058 cu. ft.)	16.387 cubic centimeters (cc; cm^3)
1 cubic inch (0.00058 cu. ft.)	0.016387 liters (l)
1 cubic foot (1,728 cu. in.)	0.028317 cubic meters (m^3)
1 cubic yard (27 cu. ft.)	0.76455 cubic meters (m^3)
1 cubic mile (cu. mi.)	4.16818 cubic kilometers (km^3)

U.S. System	Metric System
1 cubic centimeter (cc; cm^3)	0.061023 cubic inches (cu. in.)
1 cubic meter (m^3)	35.135 cubic feet (1.3079 cu. yd.)
1 cubic kilometer (km^3)	0.23390 cubic miles (cu. mi.)

TABLE C-7

Metric Conversions

Metric to U.S.	U.S. to Metric
Length	
millimeters × 0.04 = inches	inches × 25.4 = millimeters
centimeters × 0.39 = inches	inches × 2.54 = centimeters
meters × 3.28 = feet	feet × 3.04 = meters
meters × 1.09 = yards	yards × 0.91 = meters
kilometers × 0.6 = miles	miles × 1.6 = kilometers
Volume	
milliliters × 0.03 = fluid ounces	teaspoons × 5 = milliliters
milliliters × 0.06 = cubic inches	tablespoons × 15 = milliliters
liters × 2.1 = pints	fluid ounces × 30 = milliliters
liters × 1.06 = quarts	cups × 0.24 = liters
liters × 0.26 = gallons	pints × 0.47 = liters
cubic meters × 35.3 = cubic feet	quarts × 0.95 = liters
cubic meters × 1.3 = cubic yards	gallons × 3.8 = liters

(continued)

TABLE C-7 *(continued)*

Metric to U.S.	U.S. to Metric
Mass	
grams × 0.035 = ounces	ounces × 28 = grams
kilograms × 2.2 = pounds	pounds × 0.45 = kilograms
metric tons × 1.1 = short tons	short tons × 0.9 = metric tons
Area	
square centimeters × 0.16 = square inches	square inches × 6.5 = square centimeters
square meters × 1.2 = square yards	square yards × 0.8 = square meters
square kilometers × 0.4 = square miles	square miles × 2.6 = square kilometers
hectares (ha) × 2.5 = acres	acres × 0.4 = hectares
(The hectare is not an official SI unit, but it is permitted.)	

Appendix D

Tech Talk: Glossary of Terms

Imagine Alice from *Alice in Wonderland* trying to decipher a weird language when she was thrust into her strange world. *Jabberwocky*, *galumphing*, and *outgrabe* were as weird to Alice as scrum and metadata may be to you. But Alice didn't have a glossary. You do!

Alpha test: When developers test a product to get the kinks out before the product goes out for beta testing (which may be done by customers who are willing to be "test" subjects).

ARP (Address Resolution Protocol): A procedure for mapping a dynamic IP address to a permanent physical machine address in a local area network (LAN).

Artificial intelligence (AI): Leverages computers and machines to mimic problem-solving and decision-making capabilities of the human mind.

ASP (Application Service Provider): Manages and delivers applications to multiple entities from data centers across a wide area network.

Asynchronous Learning: Learning that doesn't take place in real time, but is available when learners need the training.

Augmented Reality (AR): Superimposes digital information into the user's real-world environment via devices such as smartphones and smart tablets.

Backend: What you see when you click on a web page is the frontend of the site. The backend is everything behind the scenes — servers, databases, and applications that make the website work.

Bugs: Coding mistakes or unwanted pieces of code that keep a website or program from working properly.

Caching: When a web browser stores recurring website assets — like images and font styles — so that the website will load faster on repeat visits from the same user.

Cloud Security Alliance (CSA): The world's leading organization dedicated to defining and raising awareness of best practices to help ensure a cloud-secure computing environment.

Clouds: Virtual spaces that exist on the Internet as storage space where people can place their digital resources such as software, applications, and files.

Collaborative Software: Enable the sharing, processing, and management of files, documents, and other data types among several users and/or systems. Can be used by two or more remote users to jointly work on a task.

Content Curation: The process social media sites use to gather and present content (articles, links, videos, images, and so on) that are relevant to a specific topic or a user's area of interest.

Content Management System (CMS): A software application that allows users to build and manage a website without having to code it from scratch or know how to code at all.

Context-Sensitive Help: Help that's available in online documents with the click of a mouse. Provides a quick means of accessing information.

Copyright: A type of intellectual property that protects original works of authorship.

COTS (Commercial Off-the-Shelf Software): Software that's ready to use when you buy it off the shelf. It requires little or no customizing.

Customer Relationship Management (CRM): Software that helps collect and manage a company's data about its customers and potential customers.

DEI or DE&I: Stands for Diversity, Equality, and Inclusion. It means all people should feel welcome and supported.

eLearning: A catch-all phrase for a structured course or learning experience delivered electronically.

Extended Reality (XR): An umbrella term that covers the various technologies that enhance our senses.

Extranet: An extension of an intranet that gives select customers, suppliers, business partners, or other outsiders access to part(s) of a company's intranet. For example, a company may give certain suppliers access to pricing information on an extranet.

Gamification: The application of game-design elements and principles in non-game contexts, such as in learning and instructions. It can also be defined as a set of activities and processes to solve problems by using or applying the characteristics of game-like elements.

HTML (Hypertext Markup Language): The standard language for documents on the web.

Impressions: Occurs each time someone sees a piece of corporate content online.

Intellectual Property (IP): A category of property that includes tangible and intangible inventions or creations. The best known types are patents, copyrights, and trademarks, each distinguished in this appendix.

Keywords: Words or phrases commonly used in search engines to look for online content

LAN (Local Area Network): A collection of devices connected in one physical location, such as a building office or home

LMS (Learning Management System): A digital learning environment that manages all aspects of a company's training efforts.

Metadata: Simply stated, it's data about data. It enriches the data with information that makes it easier to find on the Internet, YouTube, or more.

Metaverse technology: Refers to virtual, digital, and 3D by merging virtual spaces and the physical worlds.

Microcopy: Tiny tidbits of copy found on websites, applications, and products. They address user concerns, provide a context to situations, and create an overall positive user experience (UX).

Microlearning: Small learning units and short-term learning activities.

Mixed reality (MR): A combination of AR and VR.

Modular Chunk: A topic that stands alone for users to access only when they need the information.

Online Document: Documentation delivered on an electronic device, rather than on paper. These are often in the form of PDFs.

Organic: In marketing, the term "organic" refers to content individuals have viewed because they came to it through their natural (organic) keyword searches instead of click-through advertising.

Patent: A type of intellectual property granting exclusive right for an invention.

Pay-per-click (PPC): Advertisers pay search engines every time someone clicks on their ad.

SCORM (Shareable Content Object Reference Model): Standardizes the way eLearning courses are created and launched.

Scrum (System Customer Resolution Unraveling Meeting): A basic framework to help teams work together.

Search Engine Optimization (SEO): Techniques that improve your websites' visibility and move them up the web page search results.

Simulated Learning Environment (SLE): Simulations providing learners with procedures and routines in varying degrees of realism.

Software: A program or set of instructions that tells a computer, smartphone, or tablet what to do.

SSL (Secure Sockets Layer): Protocols for establishing authenticated and encrypted links between networked computers.

Subject Matter Expert (SME): A resource person who has an in-depth knowledge of certain subject matter.

Synchronous: Learning where the facilitator and learners interact online simultaneously, such as via videoconferencing.

Total Cost of Ownership (TCO): The cost to buy something plus the cost to operate it over its useful life.

Trademark: A type of intellectual property for any word, phrase, symbol, design, or a combination that identifies goods or services.

User Interface (UI): All the parts of a website, app, computer, smartphone, and so on, that the user can manipulate and interact with. Display and touch screens, website menus, keyboards, your cursor — these are all part of a user interface.

UX (User Experience): The creation of all the copy you see, hear, or encounter when using a digital product to enhance the user experience.

Videoconferencing: A conference among people at remote locations via transmitted audio and video signals.

Virtual Reality (VR): Places users in a virtual environment that replicates real-world experiences.

VoIP (Voice over Internet Protocol): A group of technologies for delivering voice communications over the Internet. It's also called IP telephony.

(Web) Traffic: The total amount of users who visit a website. Overall web traffic is then broken into specific types of visits — such as unique visitors and total clicks.

Appendix E

Technical Writing Brief

No matter what technical writing responsibilities you have, this Technical Writing Brief is the first step in the writing process. (You'll find a full explanation in Chapter 3.) Feel free to make a copy of this brief as you work on each new project. With just a little practice, it will help you to write with confidence and competence, as well as save you lots of time and frustration.

Many participants in the technical writing workshops I facilitate start by saying, "I don't have time to fill out this out." After working with it they say, "I don't know how I ever got along without it." Many even claim to have cut writing time by 30 to 50 percent. Share it with your team; they'll thank you!

About the Document

1. Type of document
2. Presentation context (Paper? Online? Streaming? Simulation? Combination?)
3. Target date for completion

Learner Profile

4. Who are the learners?

A. Are they technical, nontechnical, or a combination?

B. Are they internal (to your company), external, or both?

C. Do you have multi-level learners?

D. If so, what percentage are there of each?

5. What do the learners *need* to know about the topic?

A. What's their level of the subject knowledge, if any?

B. How do they process information?

C. What jobs do they perform?

6. What is their attitude toward the topic? (Positive? Neutral? Negative?)

Key Issues

7. What are the key issues to convey? (A is the most important.)

A. ______________________________

B. ______________________________

C. ______________________________

D. ______________________________

E. ______________________________

Budget

8. ____________________________________

Project Team

9. Who's who on the project team?

____

____

____

____

____

Milestones

10. List the milestones plus anticipated dates.

Milestone	Date

Approval Cycle

11. What's the approval cycle? (Start from the bottom, with A being final reviewer.)

A. ______________________

B. ______________________

C. ______________________

D. ______________________

E. 

TIP

Instead of being a book ripper and tearing this page from the book, go to `www.dummies.com` and search for **Technical Writing for Dummies Cheat Sheet**. You can print it out and keep it nearby. You can also keep on a copy on your computer, tablet, and smartphone for easy reference anywhere.

Index

B

C

D

F

G

H

I

J

K

L

M

N

O

P

Q

R

S

T

U

V

W

X

Y

Z

About the Author

Having grown up in New York City, Sheryl still bears traces of her accent, which is very apparent when she says the word "coffee" (or as she pronounces it, *cough-ee*). Sheryl stayed in New York with her husband while raising two sons, and then she became a transplanted New Yorker and moved to Boston, Massachusetts.

Sheryl became a skillful communicator because she was raised by a mother who was a stickler for communicating impeccably and spelling precisely. When she was a mere four years old, her mother taught her how to spell *antidisestablishmentarianism* and *schizophrenia*. She paraded Sheryl around like a show dog, bidding her to spell those words for anyone who'd listen. The communicating impeccably part resulted in Sheryl's stunning 25+-year career as a business communications expert — although spelling those words didn't yield her anything.

Sheryl's career has spanned writing technical documents, directing marketing and public relations campaigns, facilitating writing workshops, scripting video productions, and writing books. She's fiercely committed to helping people engage their learners, cutting writing time by 30-50 percent, and delivering key information at a glance with strong visual impact.

Writing isn't just Sheryl's craft; it's her passion. One day she hopes to write the great American novel, but in the meantime, she's written 27 books for business professionals. She also enjoys writing poetry, kayaking, cross-country skiing, snowshoeing, gardening, painting (pictures, not walls), practicing yoga, power-walking, and traveling around the world.

Sheryl Lindsell-Roberts, MA

www.linkedin.com/in/sherylwrites/

Dedication

I dedicate this book to Jon — my wonderful and patient husband. I especially appreciate Jon's patience during conversations when he perceives my focus shifting from his words to new insights into whatever book or project I'm in the midst of. Everyone needs one person who loves them for who they are and helps them to know that dreams can come true. To me, Jon is that special person.

Acknowledgments

My heartfelt thanks to my family (blood and extended) and to my dear friends. Without their love and support, I wouldn't be the person I am today — and I wouldn't be realizing my dreams.

And a loud shout out to the awesome "Dummies" (I say that with utmost respect for the real "Smarties") who made this second edition a reality. This is especially true of Tracy Boggier, Senior Acquisitions Editor, whom I thank for her steadfastness and confidence in me. I also appreciate Kezia Endsley, Project Coordinator and Editor, for her keen insights, sound advice, and for shepherding me through the *Dummies* process. Also Meir Zimmerman, who rounded out this A-team as Technical Editor, by providing expert input that kept me on my *tech toes*.

Publisher's Acknowledgments

Acquisitions Editor: Tracy Boggier
Managing Editor: Kristie Pyles
Copy Editor: Kezia Endsley
Technical Editor: Meir Zimmerman
Project Coordinator: Kezia Endsley
Production Editor: Tamilmani Varadharaj
Cover Image: © Gorodenkoff/Adobe Stock Photos